2016 SQA Past Papers With Answers

Higher
GEOGRAPHY

Higher GEOGRAPHY

2014 Specimen Question Paper, 2015 & 2016 Exam

HODDER
GIBSON
AN HACHETTE UK COMPANY

This book contains the official 2014 SQA Specimen Question Paper, 2015 and 2016 Exams for Higher Geography, with associated SQA-approved answers modified from the official marking instructions that accompany the paper.

In addition the book contains study skills advice. This advice has been specially commissioned by Hodder Gibson, and has been written by experienced senior teachers and examiners in line with the new Higher for CfE syllabus and assessment outlines. This is not SQA material but has been devised to provide further guidance for Higher examinations.

Hodder Gibson is grateful to the copyright holders, as credited on the final page of the Answer Section, for permission to use their material. Every effort has been made to trace the copyright holders and to obtain their permission for the use of copyright material. Hodder Gibson will be happy to receive information allowing us to rectify any error or omission in future editions.

Hachette UK's policy is to use papers that are natural, renewable and recyclable products and made from wood grown in sustainable forests. The logging and manufacturing processes are expected to conform to the environmental regulations of the country of origin.

Orders: please contact Bookpoint Ltd, 130 Park Drive, Milton Park, Abingdon, Oxon OX14 4SE. Telephone: (44) 01235 827720. Fax: (44) 01235 400454. Lines are open 9.00–5.00, Monday to Saturday, with a 24-hour message answering service. Visit our website at www.hoddereducation.co.uk. Hodder Gibson can be contacted direct on: Tel: 0141 333 4650; Fax: 0141 404 8188; email: hoddergibson@hodder.co.uk

This collection first published in 2016 by
Hodder Gibson, an imprint of Hodder Education,
An Hachette UK Company
211 St Vincent Street
Glasgow G2 5QY

Typeset by Aptara, Inc.

Printed in the UK

A catalogue record for this title is available from the British Library

ISBN: 978-1-4718-9089-5

3 2 1

2017 2016

Introduction

Study Skills – what you need to know to pass exams!

Pause for thought

Many students might skip quickly through a page like this. After all, we all know how to revise. Do you really though?

Think about this:

"IF YOU ALWAYS DO WHAT YOU ALWAYS DO, YOU WILL ALWAYS GET WHAT YOU HAVE ALWAYS GOT."

Do you like the grades you get? Do you want to do better? If you get full marks in your assessment, then that's great! Change nothing! This section is just to help you get that little bit better than you already are.

There are two main parts to the advice on offer here. The first part highlights fairly obvious things but which are also very important. The second part makes suggestions about revision that you might not have thought about but which WILL help you.

Part 1

DOH! It's so obvious but …

Start revising in good time

Don't leave it until the last minute – this will make you panic.

Make a revision timetable that sets out work time AND play time.

Sleep and eat!

Obvious really, and very helpful. Avoid arguments or stressful things too – even games that wind you up. You need to be fit, awake and focused!

Know your place!

Make sure you know exactly **WHEN and WHERE** your exams are.

Know your enemy!

Make sure you know what to expect in the exam.

How is the paper structured?

How much time is there for each question?

What types of question are involved?

Which topics seem to come up time and time again?

Which topics are your strongest and which are your weakest?

Are all topics compulsory or are there choices?

Learn by DOING!

There is no substitute for past papers and practice papers – they are simply essential! Tackling this collection of papers and answers is exactly the right thing to be doing as your exams approach.

Part 2

People learn in different ways. Some like low light, some bright. Some like early morning, some like evening/night. Some prefer warm, some prefer cold. But everyone uses their BRAIN and the brain works when it is active. Passive learning – sitting gazing at notes – is the most INEFFICIENT way to learn anything. Below you will find tips and ideas for making your revision more effective and maybe even more enjoyable. What follows gets your brain active, and active learning works!

Activity 1 – Stop and review

Step 1

When you have done no more than 5 minutes of revision reading STOP!

Step 2

Write a heading in your own words which sums up the topic you have been revising.

Step 3

Write a summary of what you have revised in no more than two sentences. Don't fool yourself by saying, "I know it, but I cannot put it into words". That just means you don't know it well enough. If you cannot write your summary, revise that section again, knowing that you must write a summary at the end of it. Many of you will have notebooks full of blue/black ink writing. Many of the pages will not be especially attractive or memorable so try to liven them up a bit with colour as you are reviewing and rewriting. **This is a great memory aid, and memory is the most important thing.**

Activity 2 – Use technology!

Why should everything be written down? Have you thought about 'mental' maps, diagrams, cartoons and colour to help you learn? And rather than write down notes, why not record your revision material?

What about having a text message revision session with friends? Keep in touch with them to find out how and what they are revising and share ideas and questions.

Why not make a video diary where you tell the camera what you are doing, what you think you have learned and what you still have to do? No one has to see or hear it, but the process of having to organise your thoughts in a formal way to explain something is a very important learning practice.

Be sure to make use of electronic files. You could begin to summarise your class notes. Your typing might be slow, but it will get faster and the typed notes will be easier to read than the scribbles in your class notes. Try to add different fonts and colours to make your work stand out. You can easily Google relevant pictures, cartoons and diagrams which you can copy and paste to make your work more attractive and **MEMORABLE**.

Activity 3 – This is it. Do this and you will know lots!

Step 1

In this task you must be very honest with yourself! Find the SQA syllabus for your subject (www.sqa.org.uk). Look at how it is broken down into main topics called MANDATORY knowledge. That means stuff you MUST know.

Step 2

BEFORE you do ANY revision on this topic, write a list of everything that you already know about the subject. It might be quite a long list but you only need to write it once. It shows you all the information that is already in your long-term memory so you know what parts you do not need to revise!

Step 3

Pick a chapter or section from your book or revision notes. Choose a fairly large section or a whole chapter to get the most out of this activity.

With a buddy, use Skype, Facetime, Twitter or any other communication you have, to play the game 'If this is the answer, what is the question?'. For example, if you are revising Geography and the answer you provide is "meander", your buddy would have to make up a question like "What is the word that describes a feature of a river where it flows slowly and bends often from side to side?".

Make up 10 "answers" based on the content of the chapter or section you are using. Give this to your buddy to solve while you solve theirs.

Step 4

Construct a wordsearch of at least 10 × 10 squares. You can make it as big as you like but keep it realistic. Work together with a group of friends. Many apps allow you to make wordsearch puzzles online. The words and phrases can go in any direction and phrases can be split. Your puzzle must only contain facts linked to the topic you are revising. Your task is to find 10 bits of information to hide in your puzzle, but you must not repeat information that you used in Step 3. DO NOT show where the words are. Fill up empty squares with random letters. Remember to keep a note of where your answers are hidden but do not show your friends. When you have a complete puzzle, exchange it with a friend to solve each other's puzzle.

Step 5

Now make up 10 questions (not "answers" this time) based on the same chapter used in the previous two tasks. Again, you must find NEW information that you have not yet used. Now it's getting hard to find that new information! Again, give your questions to a friend to answer.

Step 6

As you have been doing the puzzles, your brain has been actively searching for new information. Now write a NEW LIST that contains only the new information you have discovered when doing the puzzles. Your new list is the one to look at repeatedly for short bursts over the next few days. Try to remember more and more of it without looking at it. After a few days, you should be able to add words from your second list to your first list as you increase the information in your long-term memory.

FINALLY! Be inspired...

Make a list of different revision ideas and beside each one write **THINGS I HAVE** tried, **THINGS I WILL** try and **THINGS I MIGHT** try. Don't be scared of trying something new.

And remember – "FAIL TO PREPARE AND PREPARE TO FAIL!"

Higher Geography

The Course

To gain the course award, candidates must pass the learning outcomes for all three units as well as the course assessment. The purpose of the course assessment is to assess added value of the course, confirm attainment and provide a grade.

The Exam

The course assessment is structured into Component 1, which is an externally set and marked question paper worth 60 marks, and Component 2, which is an assignment based on a research topic chosen by the candidate worth 30 marks. 90 marks are available in total.

Component 1 – The External Question Paper

The question paper is set and marked by SQA, and conducted in centres under exam conditions. Candidates are expected to complete this question paper in 2 hours and 15 minutes. The questions will be asked on a local, regional and global scale and the paper has four sections.

- **Section 1: Physical Environments – 15 marks**
  Candidates must answer all questions in this section.

- **Section 2: Human Environments –15 marks**
  Candidates must answer all questions in this section.

- **Section 3: Global Issues – 20 marks**
  Candidates must answer **two** questions from out of **five** options. Candidates must answer all parts of the question within each chosen option.

- **Section 4: A question on the application of Geographical Skills – 10 marks**
  This section consists of an extended response question requiring the learner to apply geographical skills acquired during the course. These may include interpreting an Ordnance Survey map, using six-figure grid references, understanding/using scale, direction and distance. You will also be required to extract and interpret information from a variety of sources. Candidates must answer this question.

The question paper component of the course assessment will have **a greater emphasis on the assessment of knowledge and understanding** than the assignment. The other marks will be awarded for the demonstration of skills.

Component 2 – The Assignment

The purpose of the assignment is to show challenge and application by demonstrating skills, knowledge and understanding within the context of a geographical topic or issue. Candidates can choose the topic or issue to be researched.

This assignment provides candidates with the opportunity to demonstrate higher-order cognitive skills and knowledge of methods and techniques.

- The assignment involves identifying a geographical topic or issue.
- It also involves carrying out research, which should include fieldwork where appropriate.
- Candidates will be asked to demonstrate knowledge of the suitability of the methods and/or reliability of the sources used.
- The assignment will involve drawing on detailed knowledge and understanding of the topic or issue.
- It will also involve analysing and processing information from a range of sources.
- It will require reaching a conclusion supported by a range of evidence on a geographical topic or issue.
- Candidates should demonstrate the skill of communicating information.

The assignment is set by centres within SQA guidelines, and the assessment is conducted under a high level of supervision and control by the presenting centre. The production of evidence for assessment will be conducted within 1 hour and 30 minutes and with the use of specified resources. This evidence is then submitted to SQA for external marking.

The assignment component of the course assessment will have **a greater emphasis on the assessment of skills** than the question paper.

Layout of this Book

This book contains three SQA exam papers. The layout, paper colour and question level are all similar to the actual exam that you will sit, so that you are familiar with what the exam paper will look like.

The answer section is at the back of the book. Each answer contains a worked-out answer or solution so that you can see how to arrive at the right answer. The answers also include practical tips on how to tackle certain types of questions, details of how marks are awarded and advice on just what the examiners will be looking for.

As well as your class notes and textbooks, these exam papers are a useful revision tool because they will help you to get used to answering exam-style questions. You may find as you work through the questions that they refer to a case study or an example that you haven't come across before. Don't worry! You should be able to transfer your knowledge of a topic or theme to a new example. The enhanced answer section at the back will demonstrate how to read and interpret the question to identify the topic being examined and how to apply your course knowledge in order to answer the question successfully.

Examination Hints

- Make sure that you have read the instructions in the question carefully and that you have avoided needless errors such as answering the wrong sections, failing to explain when asked to or perhaps omitting to refer to a named area or case study.
- If you are asked for a named country or city, make sure you include details of any case study you have covered.
- Avoid vague answers when asked for detail. For example, avoid vague terms such as "dry soils" or "fertile soils". Instead, try to provide more detailed information in your answer such as "deep and well-drained soils" or "rich in nutrients".
- If you are given data in the form of maps, diagrams and tables in the question, make sure you refer to this information in your answer to support any points of view that you give.

- Be guided by the number of marks for a question as to the length of your answer.
- Make sure that you leave yourself sufficient time to answer all of the questions.
- One technique which you might find helpful, especially when answering long questions, is to "brainstorm" possible points for your answer. You can write these down in a list at the start of your answer and cross them out as you go through them.
- If you have any time left in the exam, use it to go back over your answers to see if you can add anything to what you have written by way of additional text or including more examples or diagrams which you may have omitted.

Common Errors

Lack of sufficient detail
- This often occurs in Higher case study answers, especially in questions with high marks.
- Many candidates fail to provide sufficient detail in answers, often by omitting reference to specific examples, or by failing to elaborate or develop points made in their answer.
- Remember that you have to give more information in your answers to gain a mark.

Irrelevant answers
- You must read the question instructions carefully to avoid giving answers which are irrelevant.
- For example, if asked to explain and you simply describe you will not score marks. If asked for a named example and you do not provide one, you will forfeit marks.

Statement reversals
- Occasionally, questions involve opposites. For example, some answers would say "death rates are high in developing countries due to poor health care" and then go on to say "death rates are low in developed countries due to good health care". Avoid doing this. You are simply stating the reverse of the first statement.
- A better second statement might be that "high standards of hygiene, health and education in developed countries have helped to bring about low death rates".

Repetition
- You should be careful not to repeat points already made in your answer. These will not gain any further marks. You may feel that you have written a long answer, but it could contain the same basic information repeated over and over. Unfortunately, these statements will be recognised and ignored when your paper is marked.

Listing
- If you give a simple list of points rather than fuller statements in your answer, you may lose marks, for example, in a 5-mark question you will obtain only 1 mark for a list.

Bullet points
- The same rule applies to a simple list of bullet points. However, if you give bullet points with some detailed explanation, you could achieve full marks.

Types of Questions and Command Words

In these past papers, and in the exam itself, a number of command words will be used in the different types of questions you will encounter. The command words are used to show you how you should answer a question; some words indicate that you should write more than others. If you familiarise yourself with these command words, it will help you to structure your answers more effectively. The question types to look out for are listed below.

Explain
These questions ask you to explain and give reasons, for example, "strategies" and relationships. If resources are provided in the question, make sure you refer to them in your answer. Some marks may be allowed for description but these will be quite restricted.

Analyse
This involves identifying parts and the relationships between them by showing the links between different components and related concepts, noting similarities and differences, explaining possible consequences and implications, and explaining the impact of, for example, processes of degradation, strategies adopted to control events and government policies on people and the environment.

Evaluate
This will involve making judgements on, for example, the relative success or failure of strategies and projects such as a river basin management scheme or aid programmes.

Discuss
These questions ask you to develop your thoughts, for example, on a specific project or change in specified situations. You may be asked to consider both sides of an argument and provide a range of comments on each viewpoint.

Geographical Skills
This question is designed to examine your geographical skills. You will be given an Ordnance Survey map along with a variety of other resources which might include a climate graph, information table, newspaper article, statistics and photographs. You will be given a set of conditions to follow; these should be used to help you answer the question. You should use/evaluate the information you have been given to make judgements and back up your answer with map evidence, information from the diagrams and tables, etc. Remember the diagrams and OS map are there for a purpose and contain valuable information that you should incorporate in your answer!

Good luck!

Remember that the rewards for passing Higher Geography are well worth it! Your pass will help you get the future you want for yourself. In the exam, be confident in your own ability. Watch your time and pace yourself carefully. If you're not sure how to answer a question, trust your instincts and just give it a go anyway – keep calm and don't panic! GOOD LUCK!

HIGHER

2014 Specimen
Question Paper

National
Qualifications
SPECIMEN ONLY

SQ20/H/01

Geography

Date — Not applicable

Duration — 2 hours and 15 minutes

Total marks — 60

SECTION 1 — PHYSICAL ENVIRONMENTS — 15 marks

Attempt ALL questions.

SECTION 2 — HUMAN ENVIRONMENTS — 15 marks

Attempt ALL questions.

SECTION 3 — GLOBAL ISSUES — 20 marks

Attempt TWO questions.

SECTION 4 — APPLICATION OF GEOGRAPHICAL SKILLS — 10 marks

Attempt the question.

Credit will be given for appropriately labelled sketch maps and diagrams.

Write your answers clearly in the answer booklet provided. In the answer booklet you must clearly identify the question number you are attempting.

Use **blue** or **black** ink.

Before leaving the examination room you must give your answer booklet to the Invigilator; if you do not you may lose all the marks for this paper.

MARKS

SECTION 1: PHYSICAL ENVIRONMENTS — 15 marks

Attempt ALL questions

Question 1

Corries are landscape features present in glaciated upland areas.

Explain the conditions and processes involved in the formation of a corrie.

You may wish to use an annotated diagram or diagrams.

5

Question 2

Look at Diagram Q2.

Explain how factors such as those shown in the diagram affect the formation of a brown earth soil.

6

Diagram Q2: Main factors affecting soil formation

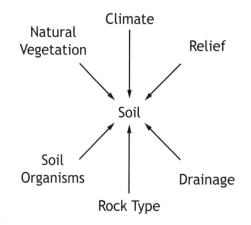

Question 3

Explain why there is a surplus of solar energy in the tropical latitudes and a deficit of solar energy towards the poles.

You may wish to use an annotated diagram or diagrams.

4

MARKS

SECTION 2: HUMAN ENVIRONMENTS — 15 marks

Attempt ALL questions

Question 1

> Nigeria conducted a population census in 2006. However, the chairperson of the National Population Commission stated in 2012 that 'Nigeria has no data. People can't really tell you precisely what the population is'. Another census will be conducted in 2016.

Explain the problems of collecting accurate population data in developing countries.

6

Question 2

> 2012 saw a significant increase in Germany's population. This was not due to a sudden baby boom, but to the many immigrants moving to the country. Experts point out this could result in both benefits and problems.

Referring to a named case study, analyse the impact of migration on **either** the donor country **or** the receiving country.

5

Question 3

Referring to **either** a named rainforest **or** a named semi-arid area, explain the techniques used to combat rural land degradation.

4

SECTION 3: GLOBAL ISSUES — 20 marks

Attempt TWO questions

MARKS

Question 1 — River Basin Management

(a)   Study Map/Data Q7 and Table Q7

Explain why there is a need for water management in Ghana.   5

Map/Data Q7

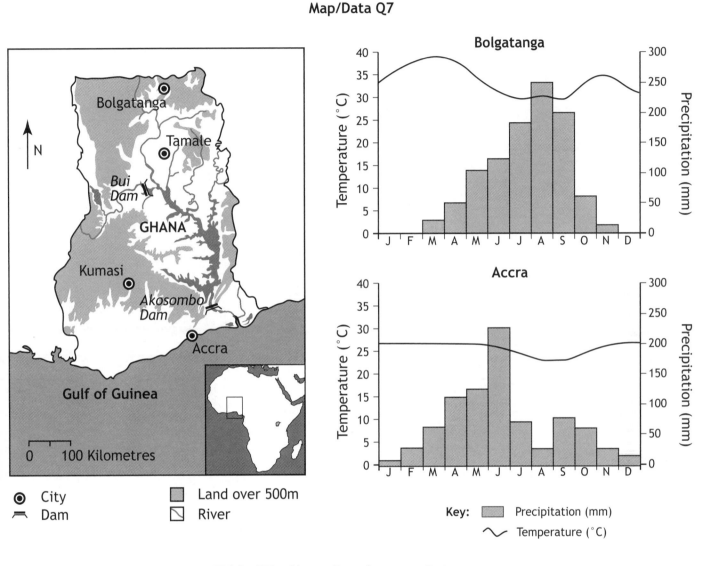

Key: Precipitation (mm)
        Temperature (°C)

Table Q7: Ghana Development Data

Current population	24·28 million
Projected population by 2040	35·8 million
Labour force by occupation	56% in agriculture 15% in industry 29% in services
% of population with access to electricity	45%

(b)   Explain the negative impacts of any named water management project.

In your answer you must refer to socio-economic **and** environmental impacts.   5

MARKS

Question 2 — Development and Health

For malaria or any other water-related disease that you have studied:

(a)   explain the methods used to try and control the spread of the disease; **and**

(b)   evaluate the effectiveness of these methods.　　　　　　　　　　　　　　**10**

MARKS

Question 3 — Global Climate Change

Look at Diagrams Q9a and Q9b.

(a) Explain the human activities which have contributed to the changes in global air temperatures.

5

Diagram Q9a: Global air temperatures 1850–2011

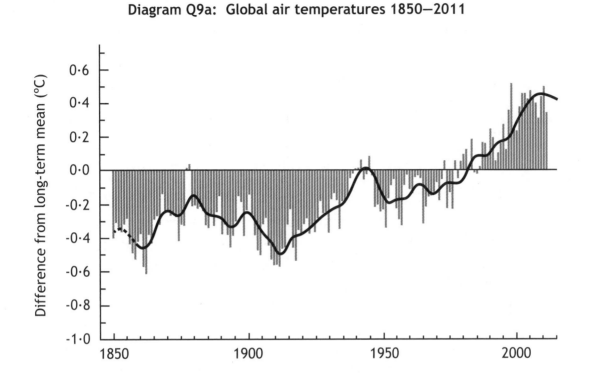

(b) Discuss the possible impacts of global warming throughout the world.

5

Diagram Q9b: Greenhouse gases, emissions by type

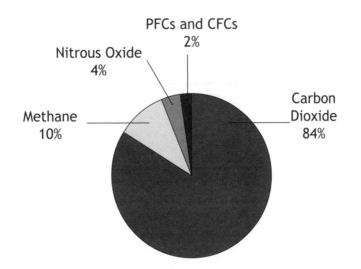

PFCs and CFCs
2%

Nitrous Oxide
4%

Methane
10%

Carbon Dioxide
84%

MARKS

Question 4 — Trade, Aid and Geopolitics

(a) Study Table Q10.

Suggest reasons for the inequalities in trade shown in the table below.　　5

Table Q10: Selected development indicators

		Germany	Thailand	Kenya
Trade statistics (million US$, 2012)	Exports	1,492,000	226,200	5,942
	Imports	1,276,000	217,800	14,390
	Balance of trade	216,000	8,400	−8,448
Economic Indicators (2012)	GDP per capita (US$) (PPP*)	39,100	1,800	10,300
	% employed in agriculture	2%	38%	75%
	% employed in manufacturing	24%	14%	10%
	% employed in services	74%	48%	5%
	Main exports	Motor vehicles, machinery, chemicals, computer and electronic products	Electronics, computer parts, automobiles and parts, electrical appliances, machinery and equipment, textiles	Tea, horticultural products, coffee, petroleum products, fish

PPP* = purchasing power parity

(b) Explain the strategies used to reduce inequalities in world trade.　　5

MARKS

Question 5 — Energy

(a) Look at Graph Q11.

Explain the differences in energy consumption between developed and developing countries.

4

Graph Q11: World energy consumption

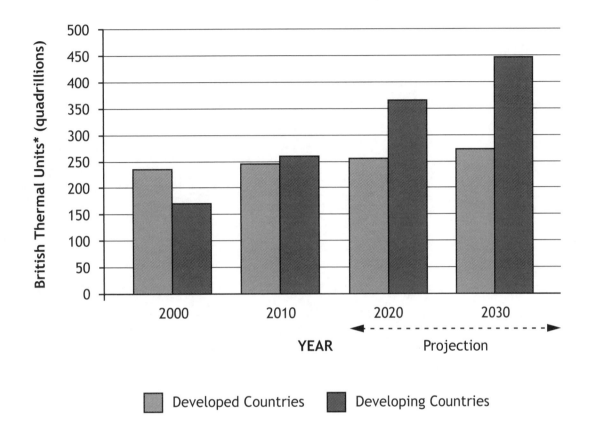

(b) Referring to different countries, evaluate the suitability of renewable approaches to generating energy.

6

OS Map (Extract 1349/EXP272: Lincoln)

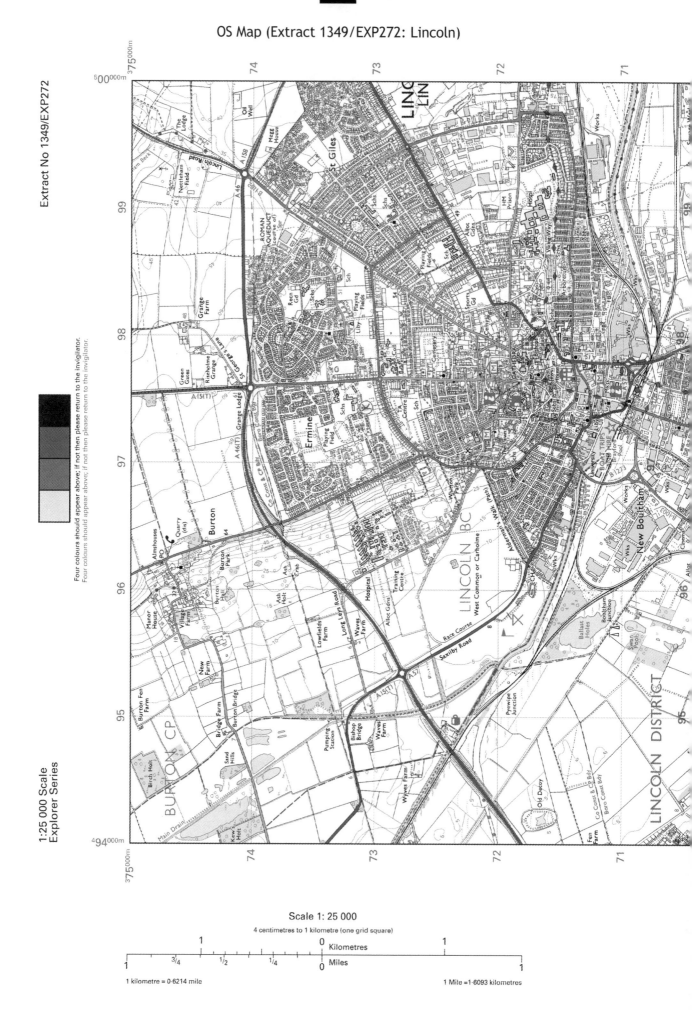

Extract No 1349/EXP272

1:25 000 Scale
Explorer Series

Scale 1: 25 000

4 centimetres to 1 kilometre (one grid square)

Kilometres

Miles

1 kilometre = 0·6214 mile

1 Mile =1·6093 kilometres

Grid North
True North
Magnetic North
Diagrammatic only

SECTION 4: APPLICATION OF GEOGRAPHICAL SKILLS — 10 marks

Attempt the question

Question 1

The city of Lincoln has decided to hold a 10 k race. Working to the brief below, a route has been proposed.

Brief for Lincoln 10 k race

The route should:

- be suitable for all participants

- cause minimum disruption to people and business in the local area

- promote business in the local area

- have a suitable start/finish line

- be scenic/interesting for participants.

Study Map Q12: Proposed 10 k Route; OS Map (Extract 1349/EXP272: Lincoln); Diagram Q12; and Graph Q12.

Referring to map evidence and other information from the sources, evaluate the suitability of the proposed route (Map Q12) in relation to the brief for the 10 k race.

You should suggest possible improvements to the route.   10

Question 1 (continued)

Map Q12: Proposed 10 k route

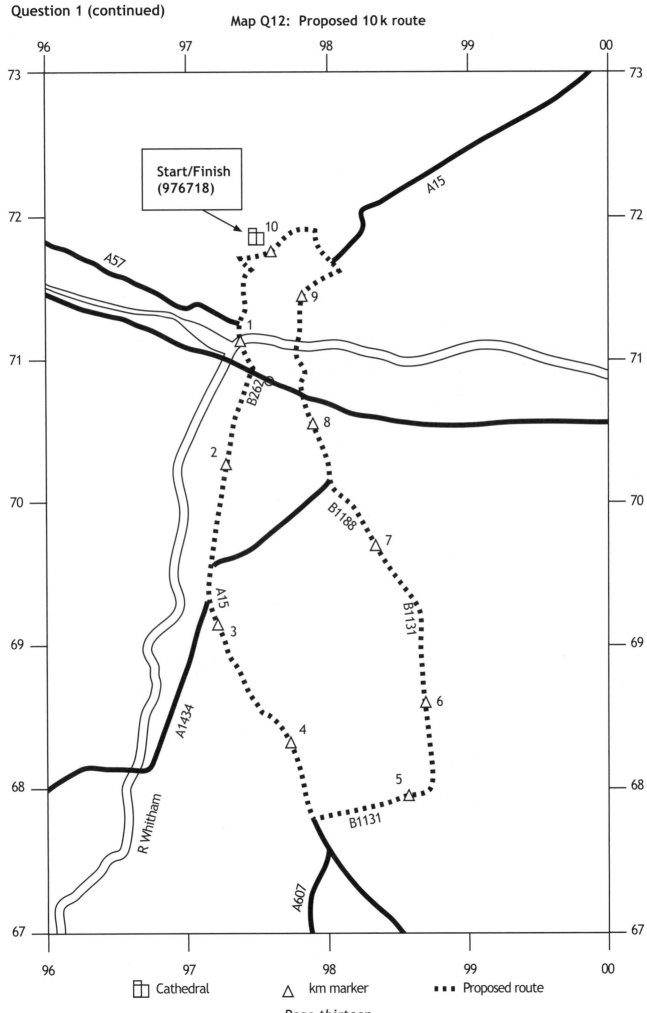

Start/Finish
(976718)

A15

A57

A15

A1434

R Whitham

A607

B262

B1188

B1131

B1131

Cathedral   km marker   ▪▪▪ Proposed route

Question 1 (continued)

Diagram Q12

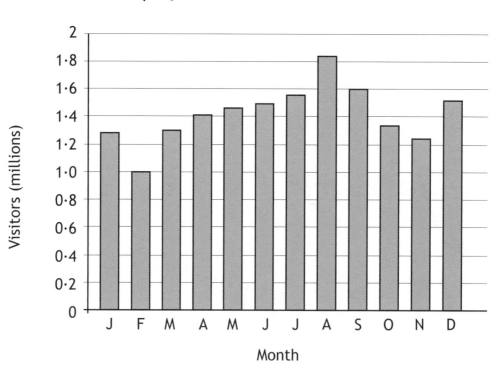

Lincoln 10 k Run
Sunday 16th February

Join over 5,000 people taking part

Live music from local bands

Safe route along closed roads

For further information on the race and nearby accommodation go to: www.visitlincoln.co.uk

Graph Q12: Visitor numbers to Lincoln

[END OF SPECIMEN QUESTION PAPER]

[BLANK PAGE]

DO NOT WRITE ON THIS PAGE

[BLANK PAGE]

DO NOT WRITE ON THIS PAGE

HIGHER

2015

National
Qualifications
2015

X733/76/11

Geography

THURSDAY, 21 MAY
9:00 AM – 11:15 AM

Total marks — 60

SECTION 1 — PHYSICAL ENVIRONMENTS — 15 marks

Attempt ALL questions.

SECTION 2 — HUMAN ENVIRONMENTS — 15 marks

Attempt ALL questions.

SECTION 3 — GLOBAL ISSUES — 20 marks

Attempt TWO questions.

SECTION 4 — APPLICATION OF GEOGRAPHICAL SKILLS — 10 marks

Attempt the question.

Credit will be given for appropriately labelled sketch maps and diagrams.

Write your answers clearly in the answer booklet provided. In the answer booklet you must clearly identify the question number you are attempting.

Use **blue** or **black** ink.

Before leaving the examination room you must give your answer booklet to the Invigilator; if you do not you may lose all the marks for this paper.

SECTION 1: PHYSICAL ENVIRONMENTS – 15 marks
Attempt ALL questions

Question 1

Study Diagram Q1 before answering this question.

Diagram Q1: Flood Hydrograph for the River Valency at Boscastle, 16 August 2004

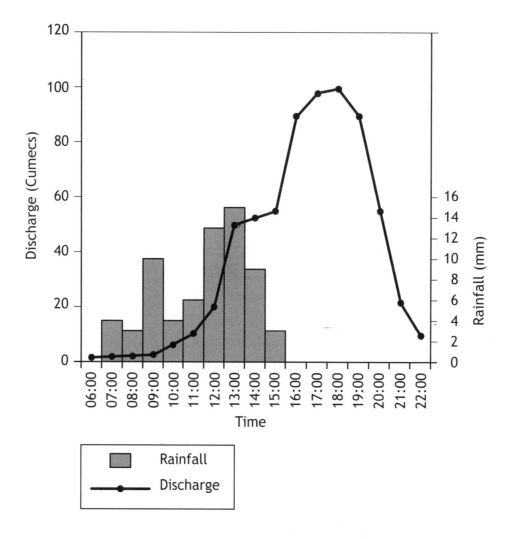

Explain the changes in discharge level of the River Valency at Boscastle on 16 August 2004. 4

MARKS

Question 2

Explain, with the aid of annotated diagrams, the various stages and processes involved in the formation of:

(a)   a stack; **and**

(b)   a sand spit.   **7**

[Turn over

Question 3

Look at Diagram Q3 before answering this question.

Diagram Q3: Surface winds and pressure zones

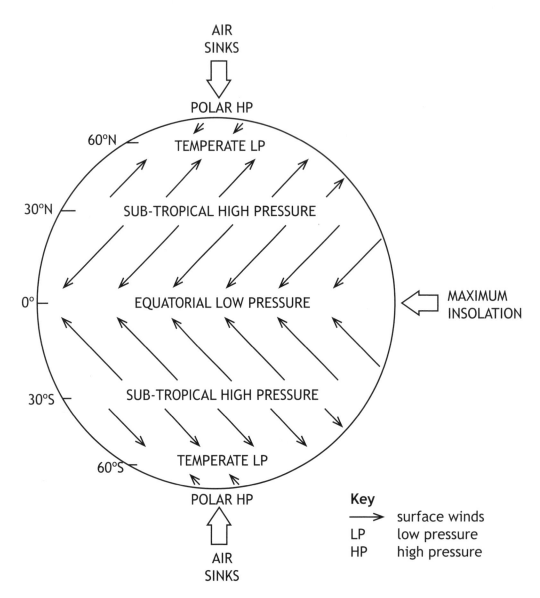

Explain how atmospheric circulation cells and the associated surface winds assist in redistributing energy around the world.　　　　4

SECTION 2: HUMAN ENVIRONMENTS – 15 marks
Attempt ALL questions

MARKS

Question 4

Study Diagrams Q4A and Q4B before answering this question.

Diagram Q4A: Population Pyramid for Ghana, 2013

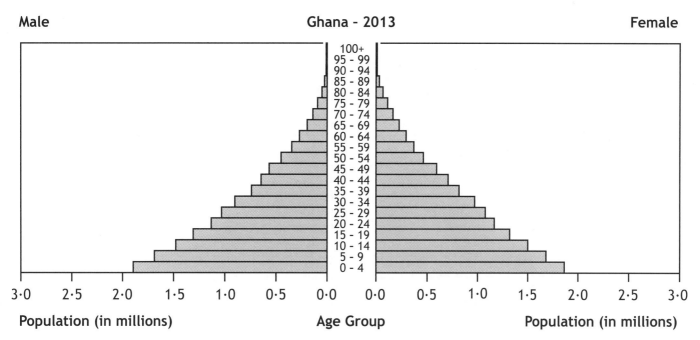

Male   Ghana – 2013   Female

Population (in millions)   Age Group   Population (in millions)

Diagram Q4B: Population Pyramid for Ghana, 2050 (predicted)

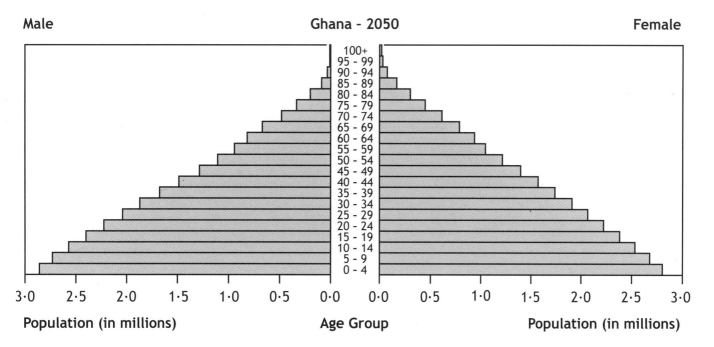

Male   Ghana – 2050   Female

Population (in millions)   Age Group   Population (in millions)

Discuss the possible consequences for Ghana of the 2050 population structure.   **5**

MARKS

Question 5

Look at Diagram Q5 before answering this question.

Traffic congestion is a major problem in cities in the UK and across the **developed** world.

Diagram Q5: Most congested UK cities, 2013

Rank	City	Congestion trend
1	Belfast	Up
2	Bristol	Level
3	Brighton	Up
4	Edinburgh	Down
5	London	Up
6	Leeds/Bradford	Down
7	Manchester	Up
8	Leicester	Up
9	Sheffield	Up
10	Liverpool	Up

Explain the strategies employed to combat the problems of traffic congestion in a **developed** world city you have studied. You should refer to specific named examples from your chosen city.

5

MARKS

Question 6

Look at Diagram Q6 before answering this question.

Rapid urbanisation in **developing** world cities has resulted in many housing problems.

Diagram Q6: Photograph of Dharavi Slum, Mumbai, India

Evaluate the impact of strategies employed to manage housing problems in a **developing** world city you have studied.

5

[Turn over

SECTION 3: GLOBAL ISSUES – 20 marks
Attempt TWO questions

MARKS

Question 7: River Basin Management

(a) **Explain** the human **and** physical factors which need to be considered when selecting a site for a major dam and its associated reservoir.   5

(b) Referring to a water control project you have studied, **explain** the **positive** social **and** economic impacts created by the construction of a major dam and its associated reservoir.   5

[Turn over

MARKS

Question 8: Development & Health

Diagram Q8: Development indicators for selected developing countries

Country	Indicators of Development					
	GDP (US $ per capita)	Employment agriculture (%)	Adult literacy (%)	Birth rate (births per 1000)	Life expectancy at birth (in years)	Hospital beds (per 1000)
Mexico	15,400	14	93	18	76	1·7
Brazil	11,700	16	90	15	73	2·3
Cuba	10,200	20	99	10	78	5·1
Kenya	1,800	75	87	30	63	1·4
Malawi	800	90	75	40	53	1·3

(a)   Study Diagram Q8 above before answering this question.

In what ways does the information in the table suggest that the five countries are at different levels of development?   4

(b)   Suggest reasons for the wide variations in development which exist between developing countries. You may wish to refer to countries that you have studied.   6

MARKS

Question 9: Global Climate Change

Look at Diagram Q9.

Diagram Q9: Natural and Enhanced Greenhouse Effect

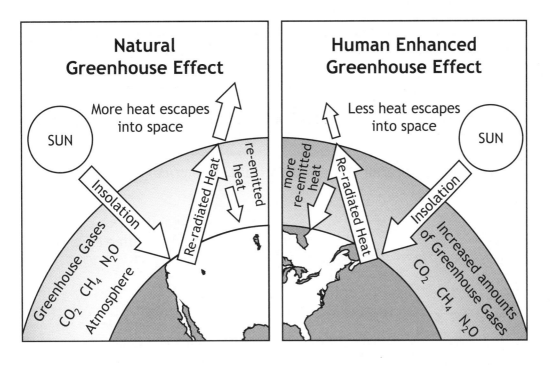

Many scientists believe that human activity has led to an enhanced greenhouse effect.

(a) **Explain** the human factors that may lead to climate change.   4

(b) **Discuss** a range of possible effects of climate change. You should support your answer with specific examples.   6

[Turn over

MARKS

Question 10: Trade, Aid and Geopolitics

Read the quotation in Diagram Q10a below.

Diagram Q10a: Quotation from Nelson Mandela (22nd November 2000)

"Where globalisation means, as it so often does, that the rich and powerful now have new means to further enrich and empower themselves at the cost of the poorer and weaker, we have a responsibility to protest in the name of universal freedom."

(a)   **Explain** the social and economic impacts of unfair trade on people and countries in the **Developing World**.   4

Diagram Q10b: Fair Trade and Normal Coffee prices

(b)   Study Diagram Q10b.

Using the information in Diagram Q10b, and also your knowledge of fair trade, **explain** how fair trade:

(i) helps to reduce inequalities in world trade; and

(ii) impacts on farming communities.   6

MARKS

Question 11: Energy

Diagram Q11: % Energy Production in Selected Countries

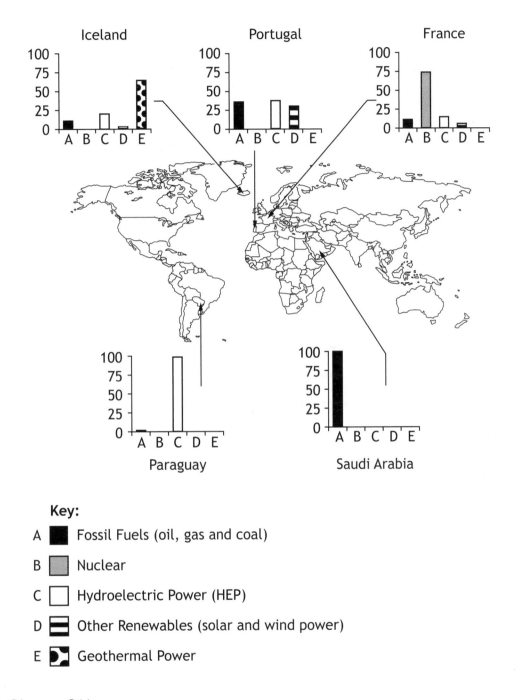

Key:

A ◼ Fossil Fuels (oil, gas and coal)

B ▨ Nuclear

C ☐ Hydroelectric Power (HEP)

D ▤ Other Renewables (solar and wind power)

E ▨ Geothermal Power

Study Diagram Q11.

(a) Using the information in the diagram **suggest reasons** for the different patterns of energy production in the countries shown.   4

(b) Choose **one** renewable *and* **one** non-renewable approach to energy production, and for each approach, **evaluate** its effectiveness in meeting energy demands.   6

SECTION 4 : APPLICATION OF GEOGRAPHICAL SKILLS – 10 marks

Attempt the question

MARKS

Question 12

> There is a proposal to develop the site at 468057 on the OS Map extract as an Outdoor and Environmental Education centre at Dalmellington, East Ayrshire.
>
> The specifications required for this development are listed below.
>
> Specifications for an Outdoor & Environmental Education Centre:
>
> - offer a range of land-based activities — hillwalking, orienteering, mountain biking and abseiling/rock-climbing
>
> - provide facilities for fishing and water-sports
>
> - allow opportunities for environmental education, fieldwork and conservation in the local area, eg biology and geography studies.

Study Diagram Q12a – Location of Proposed Development; OS Map (Extract 2144/EXP327: Dalmellington), Diagram Q12b, Diagram Q12c, Diagram Q12d and Diagram Q12e before answering this question.

Referring to map evidence from the OS Map extract, and other information from the sources, **discuss**:

(a) the advantages **and** disadvantages of the proposed location; **and**

(b) any possible impacts on the local area and East Ayrshire.

10

Diagram Q12a : Location of Proposed Development

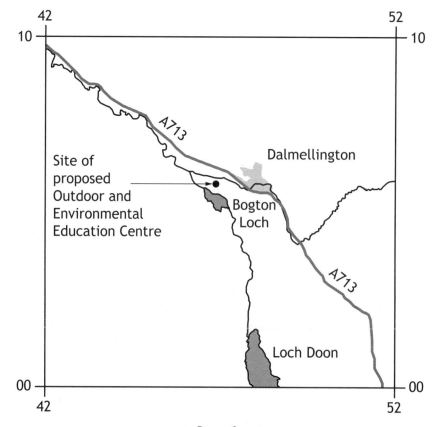

Question 12 (continued)   MARKS

Diagram Q12b : Photograph of Bogton Loch & Dalmellington Moss looking SW from GR 466065

Dalmellington Moss is a Scottish Wildlife Trust Nature Reserve. Rare heathers and moss along with birds such as curlew and snipe can be found here.

Auchenroy Hill

Bogton Loch is a Site of Special Scientific Interest because of the flora and fauna, including water birds, such as Teal and Reed Warblers, the Whooper swan and Greylag goose.

Diagram Q12c: Unemployment rates (%) for East Ayrshire and Scotland, 2008 – 2010

	2008	2009	2010
East Ayrshire	5·6	8·0	9·7
Scotland	4·5	5·9	7·6

Diagram Q12d: Population Change for Dalmellington and Burnton 2001 – 2010

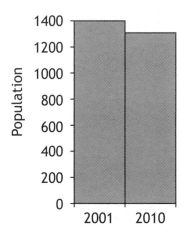

Diagram Q12e: Where tourists spend their money in Ayrshire and Arran

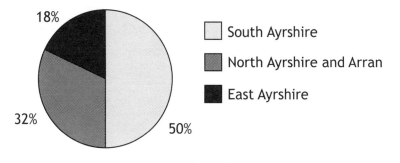

- South Ayrshire
- North Ayrshire and Arran
- East Ayrshire

18%
32%
50%

[END OF QUESTION PAPER]

[BLANK PAGE]

DO NOT WRITE ON THIS PAGE

 National
Qualifications
2015

X733/76/21

Geography
Ordnance Survey Map

THURSDAY, 21 MAY

9:00 AM – 11:15 AM

ORDNANCE SURVEY MAP

For Question 12

Note: The colours used in the printing of this map extract are indicated in the four little boxes at the top of the map extract. Each box should contain a colour; if any does not, the map is incomplete and should be returned to the Invigilator.

OS Ordnance Survey

1:25 000 Scale
Explorer Series

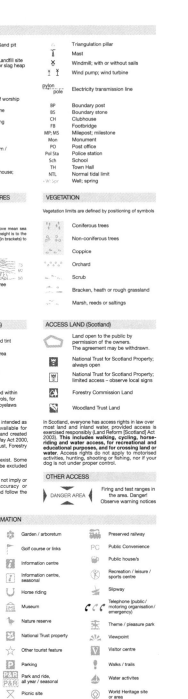

ROADS AND PATHS   Not necessarily rights of way

| M1 or A6(M) | Motorway | ⬡ Service Area | 7 Junction Number |

A 35 — Dual carriageway
A 30 — Main road
B 3074 — Secondary road
— Narrow road with passing places
— Road under construction
— Road generally more than 4 m wide
— Road generally less than 4 m wide
— Other road, drive or track, fenced and unfenced
»»» Gradient: steeper than 20% (1 in 5)
14% (1 in 7) to 20% (1 in 5)
Ferry (V) Vehicle; (P) Passenger
.... Path

RAILWAYS

Multiple track } Standard gauge
Single track
Narrow gauge or Light Rapid Transit System (LRTS) and station
Road over; road under; level crossing
Cutting; tunnel; embankment
Station, open to passengers; siding

PUBLIC RIGHTS OF WAY   Not shown on maps of Scotland

- - - - - - - Footpath
— — — Bridleway
+ + + + + Byway open to all traffic
—·—·—· Restricted byway

The representation on this map of any other road, track or path
is no evidence of the existence of a right of way

OTHER PUBLIC ACCESS

• • • Other routes with public access
◆ ◆ National Trail / Long Distance Route;
Recreational route
- - - - - Permitted footpath
—·—·— Permitted bridleway } See note below
Footpaths and bridleways along which landowners
have permitted public use but which are not rights
of way. The agreement may be withdrawn.
• • • Traffic-free cycle route
[1] National cycle network route number – traffic free
[1] National cycle network route number – on road

BOUNDARIES

—·+·— National
—·—·— County (England)
— — — Unitary Authority (UA),
Metropolitan District (Met Dist),
London Borough (LB) or District
(Scotland & Wales are solely Unitary Authorities)
············ Civil Parish (CP) (England)
or Community (C) (Wales)
▨▨▨ National Park

HISTORICAL FEATURES

✛ Site of antiquity
⚔ 1066 Site of battle (with date)
VILLA Roman
𝕮𝖆𝖘𝖙𝖑𝖊 Non-Roman
⌖ ▥▥ Visible earthwork

Magnetic North / Grid North / True North

Diagrammatic
only

GENERAL FEATURES

◉ Gravel pit   ◯ Sand pit
◔ Other pit   ◯ Landfill site
or quarry   or slag heap
⠿⠿⠿⠿ Slopes
+ Place of worship
♦ Current or former place of worship
– with tower
– with spire, minaret or dome
▢ ▭ Building; important building
▪ Glasshouse
▲ Youth hostel
▪ Bunkhouse / camping barn /
other hostel
⬟ Bus or coach station
♨ ♨ Lighthouse; disused lighthouse;
⋏ Beacon

HEIGHTS AND NATURAL FEATURES

52 · Ground survey height
·264 Air survey height

Surface heights are to the nearest metre above mean sea
level. Where two heights are shown, the first height is to the
base of the triangulation pillar and the second (in brackets) to
the highest natural point of the hill

Vertical face/cliff

Boulders   Loose   Outcrop   Scree
rock

Water; mud
Sand; sand and shingle

ACCESS LAND (England & Wales)

⬠ Access land boundary and tint
▨ Access land in wooded area
ⓘ Access information point

MANAGED ACCESS ◣ Access permitted within
managed controls, for
example, local byelaws

Portrayal of access land on this map is intended as
a guide to land which is normally available for
access on foot, for example access land created
under the Countryside and Rights of Way Act 2000,
and land managed by the National Trust, Forestry
Commission and Woodland Trust.

Access for other activities may also exist. Some
restrictions will apply; some land will be excluded
from open access rights.

The depiction of rights of access does not imply or
express any warranty as to its accuracy or
completeness. Observe local signs and follow the
Countryside Code.

TOURIST AND LEISURE INFORMATION

⌂ Building of historic interest	❀ Garden / arboretum	🚂 Preserved railway
Cadw (Welsh heritage)	Golf course or links	PC Public Convenience
⛺ Camp site	ℹ Information centre	🍺 Public house/s
Caravan site	i Information centre, seasonal	Recreation / leisure / sports centre
Camping and caravan site	∪ Horse riding	Slipway
Castle / fort	⛬ Museum	C C C Telephone (public/ motoring organisation/ emergency)
† Cathedral / Abbey	Nature reserve	Theme / pleasure park
Country park	National Trust property	Viewpoint
Cycle trail	☆ Other tourist feature	V Visitor centre
English Heritage property	P Parking	! Walks / trails
Fishing	P&R Park and ride, all year / seasonal	Water activites
Forestry Commission visitor centre	Picnic site	World Heritage site or area

GENERAL FEATURES (continued)

△ Triangulation pillar
⊥ Mast
✖ Windmill; with or without sails
✖ Wind pump; wind turbine
pylon pole Electricity transmission line
BP Boundary post
BS Boundary stone
CH Clubhouse
FB Footbridge
MP; MS Milepost; milestone
Mon Monument
PO Post office
Pol Sta Police station
Sch School
TH Town Hall
NTL Normal tidal limit
· W Spr · Well; spring

VEGETATION

Vegetation limits are defined by positioning of symbols

♣ Coniferous trees
♣ Non-coniferous trees
♧ Coppice
○ Orchard
Scrub
Bracken, heath or rough grassland
Marsh, reeds or saltings

ACCESS LAND (Scotland)

⬠ Land open to the public by
permission of the owners.
The agreement may be withdrawn.
National Trust for Scotland Property;
always open
National Trust for Scotland Property;
limited access – observe local signs
Forestry Commission Land
Woodland Trust Land

In Scotland, everyone has access rights in law over
most land and inland water, provided access is
exercised responsibly (Land Reform [Scotland] Act
2003). **This includes walking, cycling, horse-
riding and water access, for recreational and
educational purposes, and for crossing land or
water.** Access rights do not apply to motorised
activities, hunting, shooting or fishing, nor if your
dog is not under proper control.

OTHER ACCESS

◣ DANGER AREA Firing and test ranges in
the area. Danger!
Observe warning notices

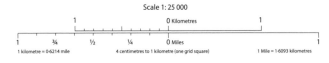

Scale 1:25 000

1 ... 0 Kilometres ... 1

1   ¾   ½   ¼   0 Miles   1
1 kilometre = 0·6214 mile   4 centimetres to 1 kilometre (one grid square)   1 Mile = 1·6093 kilometres

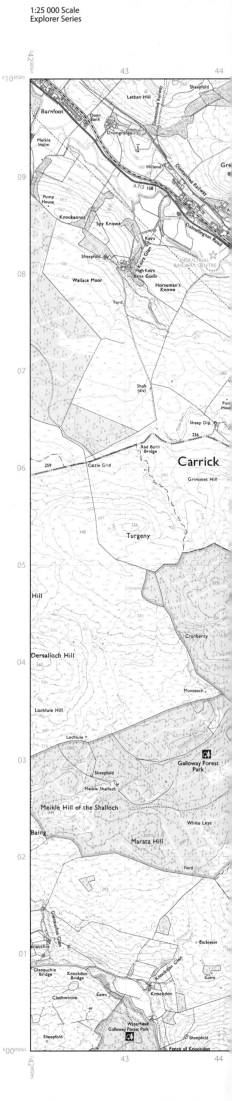

Ordnance Survey, OS, the OS Symbol and Explorer are registered trademarks of Ordnance Survey, the national mapping agency of Great Britain.
Reproduction in whole or in part by any means is prohibited without the prior written permission of Ordnance Survey. For educational use only.

Four colours should appear above; if not then please return to the invigilator.

HIGHER

2016

National
Qualifications
2016

X733/76/11

Geography

FRIDAY, 6 MAY

9:00 AM – 11:15 AM

Total marks — 60

SECTION 1 — PHYSICAL ENVIRONMENTS — 15 marks

Attempt ALL questions.

SECTION 2 — HUMAN ENVIRONMENTS — 15 marks

Attempt ALL questions.

SECTION 3 — GLOBAL ISSUES — 20 marks

Attempt TWO questions.

SECTION 4 — APPLICATION OF GEOGRAPHICAL SKILLS — 10 marks

Attempt the question.

Credit will be given for appropriately labelled sketch maps and diagrams.

Write your answers clearly in the answer booklet provided. In the answer booklet you must clearly identify the question number you are attempting.

Use **blue** or **black** ink.

Before leaving the examination room you must give your answer booklet to the Invigilator; if you do not you may lose all the marks for this paper.

MARKS

SECTION 1 — PHYSICAL ENVIRONMENTS — 15 marks
Attempt ALL questions

Question 1

Study Maps Q1A and Q1B before answering this question.

(a)   **Compare** the rainfall patterns across West Africa; **and**

(b)   **suggest reasons** for the variations.   5

Map Q1A: Location of selected air masses and the ITCZ in January and July

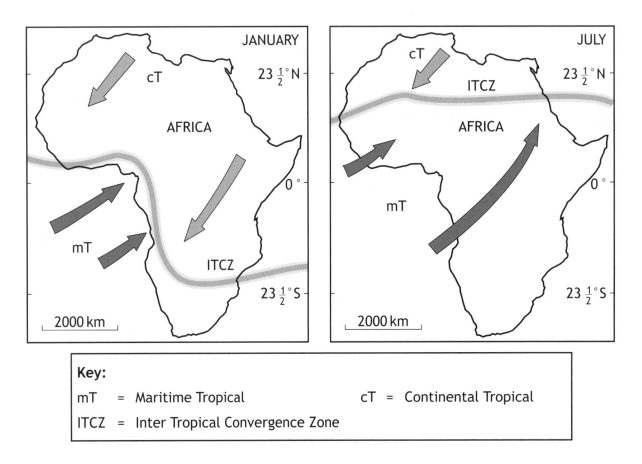

Key:

mT   =   Maritime Tropical   cT   =   Continental Tropical

ITCZ   =   Inter Tropical Convergence Zone

MARKS

Question 1 (continued)

Map Q1B: Rainfall patterns in West Africa

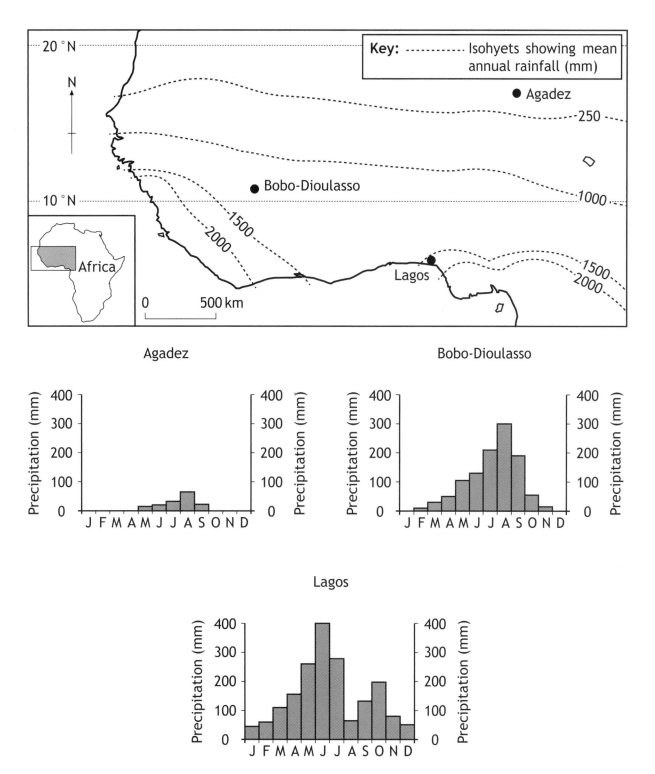

[Turn over

MARKS

Question 2

Read Diagram Q2.

Diagram Q2: Extract from Lake District Management Plan

> *The Lake District's varied scenery and historic environment provide for a wide range of sport, recreational and leisure activities . . . this has grown and changed over time.*
>
> *As the National Park continues to change, we need to monitor how access and recreation is being managed to ensure that a balance exists between the needs of everyone.*
>
> State of the Park Report 2005: Access and Recreation

Referring to a glaciated upland area you have studied:

(a) **explain** the environmental conflicts caused by the various land uses   ;   5

(b) (i) **explain** the strategies used to manage these conflicts; and

   (ii) **comment on** the effectiveness of these strategies.   5

MARKS

SECTION 2 – HUMAN ENVIRONMENTS – 15 marks
Attempt ALL questions

Question 3

Look at Map Q3 before answering this question.

The Great Green Wall Initiative aims to reduce the impact of land degradation in the Sahel zone of Northern Africa.

(a) **Explain** techniques employed to manage rural land degradation in a rainforest **or** semi arid area that you have studied; **and**

(b) **comment on** the effectiveness of these techniques.

6

Map Q3: The Great Green Wall Initiative

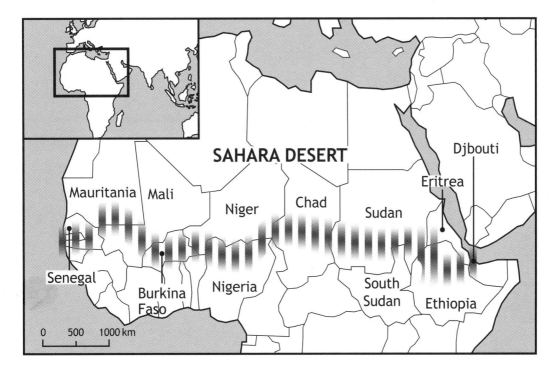

▮▮▮ "Great Green Wall" of trees

Total distance: 7,775 km

Total area: 11,662,500 hectares

[Turn over

MARKS

Question 4

> In November 2013, the Chairperson of Nigeria's National Population Commission resigned after questioning the accuracy of the data gathered about the country's population.

(a) **Discuss** how countries gather accurate population data. 3

(b) **Explain** why it is difficult to gather accurate population data in developing world countries. 6

MARKS

SECTION 3 — GLOBAL ISSUES — 20 marks
Attempt TWO questions

[Turn over

Question 5: River Basin Management

Map Q5A: Bangladesh

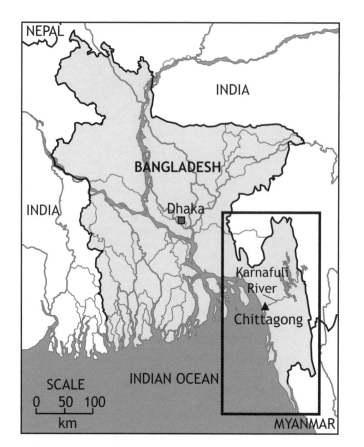

Map Q5B: Karnafuli River Basin

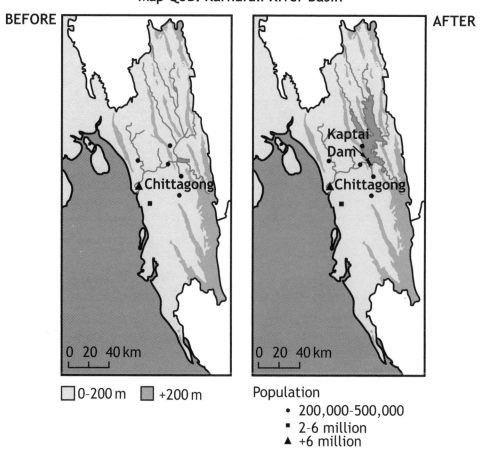

0-200 m   +200 m

Population
- • 200,000-500,000
- ▪ 2-6 million
- ▲ +6 million

MARKS

Question 5: (continued)

Diagram Q5A: Climate Graph, Chittagong, Bangladesh

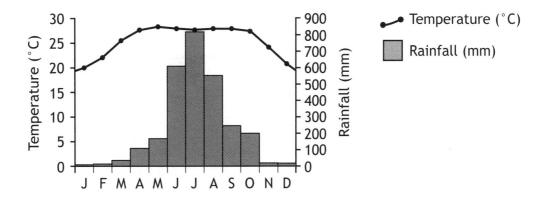

Diagram Q5B: Population growth in Bangladesh

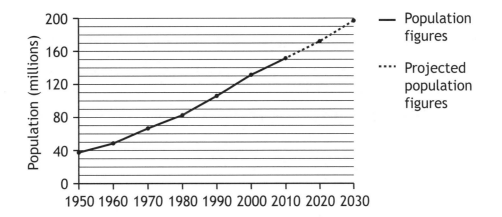

Diagram Q5C: Selected statistics, Bangladesh

Access to Improved drinking water	84%
Access to Improved sanitation	55%
Access to Electricity	62%
Main Industries	Textiles; Agriculture (rice/jute); Construction

(a)   Study Map Q5A, Map Q5B, Diagram Q5A, Diagram Q5B, and Diagram Q5C.

 Explain why there is a need for water management in the Karnafuli River Basin.   **6**

(b)   Referring to a water control project you have studied, **discuss** the positive **and** negative **environmental** impacts created by the construction of a major dam and its associated reservoir.   **4**

MARKS

Question 6: Development and Health

Look at Table Q6.

Table Q6: Selected Developing Countries

Country	GNI (Gross National Income) per capita (US$)
Venezuela	13,600
Nigeria	2,800
Malawi	900

(a) **Explain** why using only **one** development indicator, such as Gross National Income (GNI) per capita, may fail to reflect accurately the true quality of life within a country.   **3**

(b) Referring to specific Primary Health Care strategies you have studied:

 (i) **explain** how these strategies meet the health care needs of the people in a developing country; **and**

 (ii) **comment on** the effectiveness of these strategies.   **7**

MARKS

Question 7: Global Climate Change

(a)   **Explain** the **physical** causes of climate change.   4

(b)   Study Diagram Q7.

   Explain possible strategies for managing climate change.   6

Diagram Q7: Levels of managing Climate Change

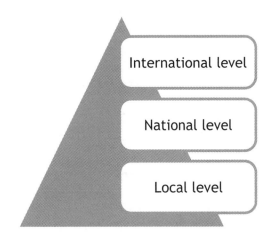

International level

National level

Local level

[Turn over

MARKS

Question 8: Trade, Aid and Geopolitics

Table Q8: Trade Patterns of Selected Countries

Country	Population (millions)	GDP per Capita (US$)	Exports (billions US$)	Imports (billions US$)	Balance of Trade (billions US$)
Australia	23	67,304	252	245	+7
USA	318	51,704	1,575	2,273	−698
Germany	81	41,866	1,493	1,233	+260
UK	64	39,161	475	646	−171
Saudi Arabia	30	24,524	376	147	+229
Botswana	2	16,400	3	7	−4
Russia	144	14,302	515	341	+174
Brazil	201	11,358	245	241	+4
South Africa	53	7,525	91	99	−8
China	1,361	6,071	2,210	1,772	+438
Ghana	25	3,500	13	18	−5
India	1,241	1,499	318	516	−198
Zimbabwe	13	600	7	4	+3

(Data from CIA World Fact book — 2013)

(a) Study Table Q8 above.

   To what extent does the data in the table show inequalities in the pattern of world trade? 4

(b) **Give reasons** for the inequalities in the pattern of trade shown in the table or between other countries that you have studied. 6

Question 9: Energy   MARKS

Graph Q9A: Total UK energy consumption (1970-2030)

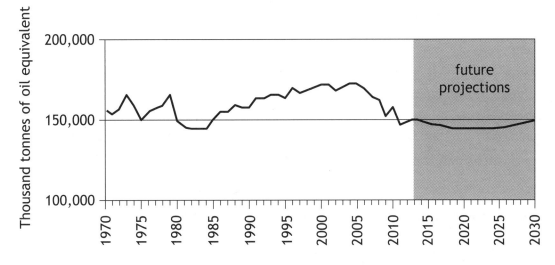

(a)   Look at Graph Q9A above.

      Suggest reasons for the changes in energy consumption in the UK.   5

Diagram Q9B: Hydraulic Fracturing ("Fracking") to extract shale gas

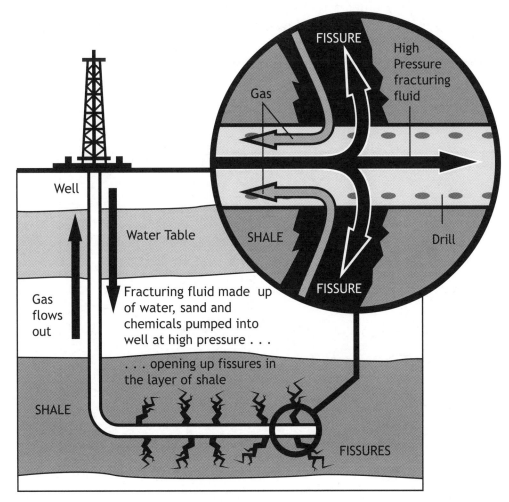

(b)   Look carefully at Diagram Q9B above.

      Discuss the advantages **and** disadvantages of hydraulic fracturing ("fracking"), or any other **non**-renewable source of energy you have studied, in meeting the energy demands of a country.   5

[Turn over

MARKS

SECTION 4 — APPLICATION OF GEOGRAPHICAL SKILLS — 10 marks
Attempt the question

Question 10

East Sussex County Council is building a new by-pass from 743091 to 776111 (The Bexhill-Hastings Link Road), however, not everyone is pleased with this decision.

Study the OS Map (Extract No 2214/EXP124: Hastings and Bexhill), Map Q10, Diagram Q10A, Diagram Q10B, Diagram Q10C and Diagram Q10D, before answering this question.

Referring to evidence from the OS map extract, and other information from the sources, **discuss**:

(a)   the advantages **and** disadvantages of the proposed route; **and**

(b)   any possible impacts on the surrounding area.   **10**

Diagram Q10A: Part of Sussex Wildlife Trust's response to the new road proposal

> "Sussex Wildlife Trust strongly objects to the Bexhill-Hastings Link Road Scheme. The proposals do not represent sustainable development. The scheme will result in unacceptable environmental damage."

Map Q10: Route of the Bexhill to Hastings Link Road

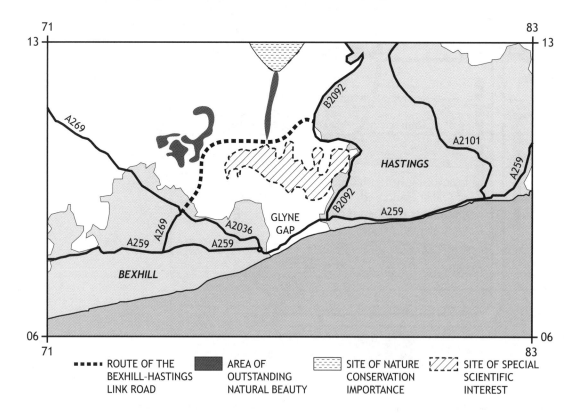

| ▪▪▪▪▪ ROUTE OF THE BEXHILL-HASTINGS LINK ROAD | AREA OF OUTSTANDING NATURAL BEAUTY | SITE OF NATURE CONSERVATION IMPORTANCE | SITE OF SPECIAL SCIENTIFIC INTEREST |

Question 10 (continued) MARKS

Diagram Q10B: Average Hourly Traffic Flow on the A259 at Glyne Gap 768082 June 2012

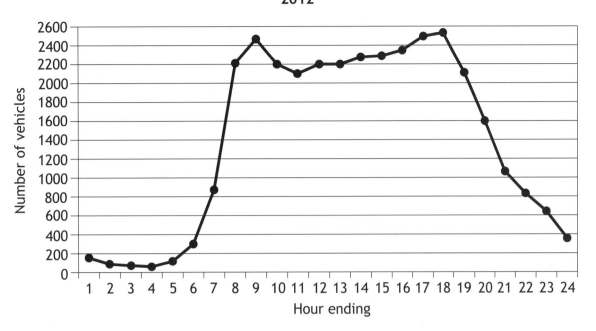

Diagram Q10C: Number of days that safe air pollution levels were exceeded at Glyne Gap 768082 June 2011 to June 2013

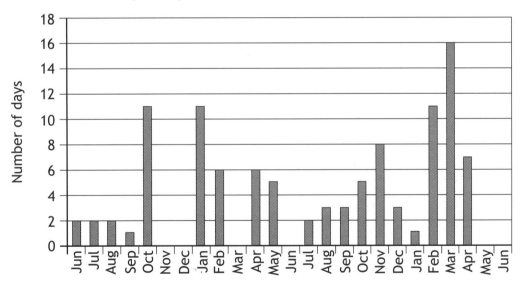

Diagram Q10D: Predicted visual impact on the Combe Havan Valley (View East from 757107) towards Adam's farm

Current View   Artist impression of view in 2020

[END OF QUESTION PAPER]

[BLANK PAGE]

DO NOT WRITE ON THIS PAGE

National
Qualifications
2016

X733/76/21

Geography
Ordnance Survey Map

FRIDAY, 6 MAY

9:00 AM – 11:15 AM

ORDNANCE SURVEY MAP

For Question 10

Note: The colours used in the printing of this map extract are indicated in the four little boxes at the top of the map extract. Each box should contain a colour; if any does not, the map is incomplete and should be returned to the Invigilator.

Ordnance Survey

1:25 000 Scale
Explorer Series

ROADS AND PATHS Not necessarily rights of way

M1 or A6(M)	Motorway
	Service Area Junction Number
A 35	Dual carriageway
A 30	Main road
B 3074	Secondary road
	Narrow road with passing places
	Road under construction
	Road generally more than 4 m wide
	Road generally less than 4 m wide
	Other road, drive or track, fenced and unfenced
	Gradient: steeper than 20% (1 in 5) 14% (1 in 7) to 20% (1 in 5)
Ferry	Ferry; Ferry P – passenger only
	Path

RAILWAYS

	Multiple track Standard gauge
	Single track
	Narrow gauge or Light Rapid Transit System (LRTS) and station
	Road over; road under; level crossing
	Cutting; tunnel; embankment
	Station, open to passengers; siding

PUBLIC RIGHTS OF WAY Not shown on maps of Scotland

- - - - - - -	Footpath
— — — — —	Bridleway
-+-+-+-+-+-	Byway open to all traffic
·+·+·+·+·	Restricted byway-not for use by mechanically propelled vehicles

The representation on this map of any other road, track or path is no evidence of the existence of a right of way

OTHER PUBLIC ACCESS

◦ ◦ ◦	Other routes with public access

The exact nature of the rights on these routes and the existence of any restrictions may be checked with the local highway authority. Alignments are based on the best information available

◆ ◆ ◆	Recreational route
◆ ◆ ◆ National Trail / (✦) Long Distance Route	
- - - - - -	Permissive footpath	See note below
— — — —	Permissive bridleway

Footpaths and bridleways along which landowners have permitted public use but which are not rights of way. The agreement may be withdrawn.

◦ ◦ ◦	Traffic-free cycle route
1 1	National cycle network route number – traffic free; on road

BOUNDARIES

—·—+—·—+—	National
— — — —	County (England)
— — — —	Unitary Authority (UA), Metropolitan District (Met Dist), London Borough (LB) or District (Scotland & Wales are solely Unitary Authorities)
············	Civil Parish (CP) (England) or Community (C) (Wales)
▭▭▭▭	National Park

HISTORICAL FEATURES

✦	Site of antiquity
⚔ 1066	Site of battle (with date)
VILLA	Roman
Castle	Non-Roman
⌂	Visible earthwork

Information provided by English Heritage for England and the Royal Commissions on the Ancient and Historical Monuments for Scotland and Wales

GENERAL FEATURES

	Gravel pit		Sand pit
	Other pit or quarry		Landfill site or slag/spoil heap
⊥⊥⊥⊥⊥	Slopes		
+	Place of worship		
	Current or former place of worship		
	– with tower		
	– with spire, minaret or dome		
▭ ▭	Building; important building		
■	Glasshouse		
▲	Youth hostel		
■	Bunkhouse / camping barn / other hostel		
⊡	Bus or coach station		
⚲ ⚲	Lighthouse; disused lighthouse;		
⟁	Beacon		

Ⅰ	Triangulation pillar
ⵋ	Mast
⚡	Windmill; with or without sails
ⵋ	Wind pump; wind turbine
pylon pole	Electricity transmission line
BP	Boundary post
BS	Boundary stone
CG	Cattle grid
CH	Clubhouse
FB	Footbridge
MP; MS	Milepost; milestone
Mon	Monument
PO	Post office
Pol Sta	Police station
Sch	School
TH	Town hall
NTL	Normal tidal limit
W; Spr	Well; spring

HEIGHTS AND NATURAL FEATURES

52	Ground survey height
·284	Air survey height

Surface heights are to the nearest metre above mean sea level. Where two heights are shown, the first height is to the base of the triangulation pillar and the second (in brackets) to the highest natural point of the hill

Vertical face/cliff

Boulders   Loose rock   Outcrop   Scree

	Water; mud
	Sand; sand and shingle

VEGETATION

Vegetation limits are defined by positioning of symbols

	Coniferous trees
	Non-coniferous trees
	Coppice
	Orchard
	Scrub
	Bracken, heath or rough grassland
	Marsh, reeds or saltings

ACCESS LAND

DANGER AREA	Firing and test ranges in the area. Danger! Observe warning notices	MANAGED ACCESS	Access permitted within managed controls, for example, local byelaws.

England and Wales

	Access land boundary and tint
	Access land in wooded area
ⓘ	Access information point

Portrayal of access land on this map is intended as a guide to land which is normally available for access on foot, for example access land created under the Countryside and Rights of Way Act 2000, and land managed by the National Trust, Forestry Commission and Woodland Trust.

Access for other activities may also exist. Some restrictions will apply; some land will be excluded from open access rights.

The depiction of rights of access does not imply or express any warranty as to its accuracy or completeness. Observe local signs and follow the Countryside Code.

Scotland

	National Trust for Scotland, always open
	National Trust for Scotland, limited access – observe local signs
	Forestry Commission Land
	Woodland Trust Land

In Scotland, everyone has access rights in law over most land and inland water, provided access is exercised responsibly (Land Reform [Scotland] Act 2003). This includes walking, cycling, horse-riding and water access, for recreational and educational purposes, and for crossing land or water. Access rights do not apply to motorised activities, hunting, shooting or fishing, nor if your dog is not under proper control.

TOURIST AND LEISURE INFORMATION

⛨	Building of historic interest	⛰	Forestry Commission visitor centre
⛵	Boat trips	❀	Garden / arboretum
⚓	Boat hire	⚐	Golf course or links
⛯	Cadw (Welsh heritage)	⌂	Historic Scotland
⛺	Camp site/Caravan site	ⓘ	Information centre, seasonal
⛫	Castle / fort	⛾	Horse riding
✝	Cathedral / Abbey	⌂	Museum
⚒	Craft Centre	⛰	Mountain bike trail
⛨	Country park	⚘	Nature reserve
⚙	Cycle hire	⛺	National Trust property
⚙	Cycle trail	☆	Other tourist feature
⛨	English Heritage property	P P&R	Parking / Park and ride, all year / seasonal
⚲	Fishing	✕	Picnic site

⛟	Preserved railway
PC	Public Convenience
⛒	Public house/s
⚘	Recreation / leisure / sports centre
⚓	Slipway
✆ ✆	Telephone (public / roadside assistance / emergency)
⛱	Theme / pleasure park
⚆	Viewpoint
V	Visitor centre
!	Walks / trails
◎	Water activites
◎	World Heritage site or area

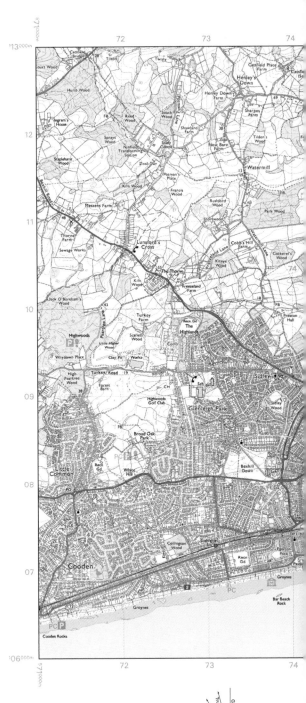

Scale 1: 25 000

1   0 Kilometres   1

1   ¾   ½   ¼   0 Miles   1

1 kilometre = 0·6214 mile   4 centimetres to 1 kilometre (one grid square)   1 Mile = 1·6093 kilometres

General Marking Principles for Higher Geography

This information is provided to help you understand the general principles you must apply when marking candidate responses to questions in this Paper. These principles must be read in conjunction with the detailed marking instructions, which identify the key features required in candidate responses.

a) Marks for each candidate response must always be assigned in line with these General Marking Principles and the Detailed Marking Instructions for this assessment.

b) Marking should always be positive. This means that, for each candidate response, marks are accumulated for the demonstration of relevant skills, knowledge and understanding: they are not deducted from a maximum on the basis of errors or omissions.

c) Where the candidate violates the rubric of the paper and answers two parts in one section, both responses should be marked and the better mark recorded.

d) Marking must be consistent. Never make a hasty judgement on a response based on length, quality of hand writing or a confused start.

e) Use the full range of marks available for each question.

f) The Detailed Marking Instructions are not an exhaustive list. Other relevant points should be credited.

g) For credit to be given, points must relate to the question asked. Where candidates give points of knowledge without specifying the context, these should be rewarded unless it is clear that they do not refer to the context of the question.

h) For knowledge/understanding marks to be awarded, points must be:
   a. relevant to the issue in the question
   b. developed (by providing additional detail, exemplification, reasons or evidence)
   c. used to respond to the demands of the question (ie evaluate, analyse, etc)

Marking principles for each question type

There are a range of types of question which could be asked within this question paper. For each, the following provides an overview of marking principles, and an example for each.

Explain

Questions which ask candidates to explain or suggest reasons for the cause or impact of something, or require them to refer to causal connections and relationships: candidates must do more than describe to gain credit here.

Where this occurs in a question asking about a landscape feature, candidates should refer to the processes leading to landscape formation.

Where candidates are provided with sources, they should make use of these and refer to them within their answer for full marks.

Where candidates provide a purely descriptive answer, or one where development is limited, no more than half marks should be awarded for the question.

Other questions look for higher-order skills to be demonstrated and will use command words such as analyse, evaluate, to what extent does, discuss.

Analyse

Analysis involves identifying parts, the relationship between them, and their relationships with the whole. It can also involve drawing out and relating implications.

An analysis mark should be awarded where a candidate uses their knowledge and understanding/a source, to identify relevant components (eg of an idea, theory, argument, etc) and clearly show at least one of the following:

- links between different components
- links between component(s) and the whole
- links between component(s) and related concepts
- similarities and contradictions
- consistency and inconsistency
- different views/interpretations
- possible consequences/implications
- the relative importance of components
- understanding of underlying order or structure

Where candidates are asked to analyse they should identify parts of a topic or issue and refer to the interrelationships between, or impacts of, various factors, eg analyse the soil-forming properties which lead to the formation of a gley soil. Candidates would be expected to refer to how the various soil formatting properties contributed to the formation.

Evaluate

Where candidates are asked to evaluate, they should be making a judgement of the success, failure, or impact of something based on criteria. Candidates would be expected to briefly describe the strategy/project being evaluated before offering an evidenced conclusion.

Account for

Where candidates are being asked to account for, they are required to give reasons, often (but not exclusively) from a resource, eg for a change in trade figures, a need for water management, or differences in development between contrasting developing countries.

Discuss

These questions are looking for candidates to explore ideas about a project, or the impact of a change. Candidates will be expected to consider different views on an issue/argument. This might not be a balanced argument, but there should be a range of impacts or ideas within the answer.

To what extent

This asks candidates to consider the impact of a management strategy or strategies they have explored. Candidates would be expected to briefly describe the strategy/project being evaluated before offering an evidenced conclusion. Candidates do not need to offer an overall opinion based on a variety of strategies, but should assess each separately.

Marking Instructions for each question

Section 1: Physical Environments

Question		General Marking Instructions for this type of question	Max mark	Specific Marking Instructions for this question
1.		Check any diagram(s) for relevant points not present in the text and award accordingly. Well-annotated diagrams that explain conditions and processes can gain full marks. Maximum of 1 mark for undeveloped conditions and 1 mark for undeveloped processes. Award a maximum of 2 marks for fully developed processes. Answers must have conditions and processes to gain maximum marks. Answers which are purely descriptive, or do not develop any processes or conditions, should achieve no more than 2 marks in total. **1 mark** Limited explanation — the use of the names of at least two processes with no development of these. **2 marks** The use of the names of at least two processes/conditions with development of these, but no other reference to conditions. **OR** Limited use of the names of at least two processes/conditions, with at least two descriptive points about the landscape formation. **3 marks** The use of the processes of at least two conditions with development of these, or two developed processes with limited explanation of how the feature forms over time. Limited use of the names of at least three processes/conditions, with at least three descriptive points about the landscape formation. **4 marks** The use of the processes of at least three processes/conditions with development of these, or two developed processes, with two further statements explaining the formation of the feature.	5	*Possible answers might include:* • Snow accumulates in mountain hollows when more snow falls in winter than melts in the summer. (1 mark) • North/north-east facing slopes are more shaded so snow lies longer (1 mark), with accumulated snow compressed into neve and eventually ice. (1 mark) • Plucking, when ice freezes on to bedrock, pulling loose rocks away from the backwall, making it steeper. (1 mark) • Abrasion, when the angular rock embedded in the ice grinds the hollow, making it deeper. (1 mark) • Frost shattering continues to steepen the sides of the hollow when water in cracks in the rock turns to ice when temperatures drop below freezing; expansion and contraction weakens the rock until fragments break off. (1 mark) • Rotational sliding further deepens the central part of the hollow floor as gravity causes the ice to move. (1 mark) • Friction causes the ice to slow down at the front edge of the corrie, allowing a rock lip to form, which traps water as ice melts, leaving a lochan or tarn. (1 mark) • During spring/summer, thawing takes place, allowing water to penetrate cracks in the rocks at the base of the hollow. (1 mark) The broken fragments build up over time and are removed by meltwater, further enlarging the hollow. (1 mark) Frost shattering on the backwall supplies further abrasion material as loose scree falls down the bergschrund. (1 mark) • This is a large crevasse separating moving ice from the ice still attached to the backwall. (1 mark) A weaker answer may take the form of descriptive points such as: • A corrie forms when a glacier forms in a hollow and moves downhill, eroding an armchair shape. (1 mark) Plucking, abrasion and frost shattering help to erode the corrie. (1 mark) The ice steeped the back wall and deepens the hollow. (1 mark)

1.		(continued)		
		5 marks The use of the processes of at least three processes/conditions with development of these, or three developed processes, with three further statements explaining the formation of the feature, including a named example.		
2.		Candidates need not refer to all factors in diagram for full marks — at least two factors are expected. "Explain" questions should make reference to *causal* relationships. Marks may be awarded as follows: • For 1 mark, candidates may give one detailed explanation, or a limited description/explanation of two factors. • Candidate responses may be a mixture of the two styles of writing, but a maximum of 3 marks should be awarded for answers entirely consisting of limited descriptive/explanatory points.	6	*Possible answers might include:* • Natural vegetation — deciduous forest vegetation provides deep leaf litter, which is broken down rapidly in mild/warm climate. (1 mark) • Trees have roots which penetrate deep into the soil, ensuring the recycling of minerals back to the vegetation. (1 mark) • Soil organisms — soil biota break down leaf litter producing mildly acidic mull humus. They also ensure the mixing of the soil, aerating it and preventing the formation of distinct layers within the soil. (1 mark) • Climate — precipitation slightly exceeds evaporation, giving downward leaching of the most soluble minerals and the possibility of an iron pan forming, impeding drainage. (1 mark) • Rock type — determines the rate of weathering, with hard rocks such as schist taking longer to weather, producing thinner soils. Softer rocks, eg shale, weather more quickly. (1 mark) • Relief — greater altitude results in temperatures and the growing season being reduced and an increase in precipitation. (1 mark) Steeper slopes tend to produce thinner soils due to gravity. (1 mark) • Drainage — well drained with throughflow and little accumulation of excess water collecting, producing limited leaching. (1 mark) Candidates may give a developed explanations with interactions between factors, for example: • The A horizon is rich in nutrients, caused by the relatively quick decomposition of the litter of deciduous leaves and grasses in a mild climate. (1 mark) • This produces a mull humus, well mixed with the soil minerals thanks to the activity of organisms such as worms. (1 mark) • Soil colour varies from black humus to dark brown in A horizon to lighter brown in B horizon where humus content is less obvious. Texture is loamy and well-aerated in the A horizon but lighter in the B horizon. (1 mark) • The C horizon is derived from a range of parent material, with limestone producing lighter-coloured alkaline soils. (1 mark) • South-facing slopes with a greater amount of sunshine and higher temperatures increase the rate of humus production. (1 mark)

| 3. | Check any diagram(s) for relevant points not present in the text and award accordingly.

Well-annotated diagrams, explaining reasons, can gain full marks.

• For 1 mark, candidates may give one detailed explanation, or a limited description/explanation of two factors.

• Candidate responses may be a mixture of the two styles of writing, but a maximum of 2 marks should be awarded for answers entirely consisting of limited descriptive/explanatory points. | 4 | *Possible answers might include:*

• Sun's angle in the sky decreases towards the poles due to the curvature of the Earth, which spreads heat energy over a larger surface area. (1 mark)
• Sun's rays are concentrated on tropical latitudes as the intensity of insolation is greatest where rays strike vertically. (1 mark)
• Sun's rays have less atmosphere to pass through at the tropics, so less energy is lost through absorption and reflection by clouds, gas and dust. (1 mark)
• Albedo rates differ from the darker forest surfaces at the tropics absorbing radiation, in contrast to the ice-/snow-covered polar areas reflecting radiation. (1 mark)
• Tilt of the axis results in the Sun being higher in the sky between the tropics throughout the year, focusing energy. (1 mark)
• No solar insolation at the winter solstices at the poles producing 24-hour darkness, whereas the tropics receive insolation throughout the year. (1 mark) |

Section 2: Human Environments

Question	General Marking Instructions for this type of question	Max mark	Specific Marking Instructions for this question
1.	Candidates must explain the problems of collecting accurate population data in developing countries. No marks for description. Marks may be awarded as follows: • For 1 mark, candidates may give one detailed explanation, or a limited description/explanation of two factors. Detail may include relevant exemplification of a problem. • Candidate responses may be a mixture of the two styles of writing, but a maximum of 3 marks should be awarded for answers entirely consisting of limited descriptive/explanatory points.	6	*Possible answers might include:* • Large numbers of migrants, eg the Tuareg or Fulani in West Africa, and the shifting cultivators of the Amazon, may lead to people being missed or counted twice. (1 mark) • Countries with large numbers of homeless people or large numbers of rural-to-urban migrants living in shanty towns, eg Makoko in Lagos, Nigeria, have no official address for an enumerator to visit. (1 mark) • Poor communication links and difficult terrain, eg in the Amazon Rainforest, may make it difficult for enumerators to reach isolated villages. (1 mark) • The variety of languages spoken in many countries (eg over 500 in Nigeria) make it difficult to provide forms that everyone can complete. (1 mark) • The considerable costs involved in printing, training enumerators, distributing forms and analysing the results can make conducting a census impossible, especially when the country may have more pressing problems like housing and education. (1 mark) • In countries with high levels of illiteracy, mistakes may be made and more enumerators will be needed to help. (1 mark) • People may be suspicious of why the census is being conducted, and may lie. (1 mark) • Ethnic tensions and internal political rivalries may lead to inaccuracies, eg northern Nigeria was reported to have inflated its population figures to secure increased political representation. (1 mark) • Under-registration may occur for social, religious and political reasons, eg China's one-child policy may have reduced the registration of baby girls. (1 mark) • In countries suffering from war, eg Afghanistan, it may be dangerous for enumerators to enter regions and data will quickly become dated. (1 mark)

| 2. | | Candidates should analyse the impact of the migration on either the donor or the recipient country. Advantages and disadvantages must be included for full credit.

Markers should take care not to credit reversed points.

Answers will depend on the case study described by the candidate.

Marks may be awarded as follows:

• For 1 mark, candidates may give one detailed explanation, or a limited description/explanation of two factors. Detail may include relevant exemplification of a problem.

• Candidate responses may be a mixture of the two styles of writing, but a maximum of 2 marks should be awarded for answers entirely consisting of limited descriptive/explanatory points.

A maximum of 2 marks should be awarded for answers which are vague or over-generalised. | 5 | *Possible answers might include:*

Donor country, eg Greece, Spain or Bulgaria:
Advantages
• Pressure on local services such as education, healthcare and housing is reduced. (1 mark)
• Pressure on jobs is reduced therefore levels of unemployment will fall. (1 mark)
• The birth rate is lowered so population growth rates will slow. (1 mark)
• Money sent home by the migrants will boost the local economy. (1 mark)
• Migrants will learn new skills and may then return to their home country. (1 mark)

Disadvantages
• Active and most educated population left, known as the 'brain drain', which resulted in a skills shortage in donor countries. (1 mark)
• Families were divided and the death rates may increase due to the ageing population. (1 mark)
• Family members remaining in the country of origin may become dependent on remittances being sent home by migrant workers. (1 mark)

Recipient country, eg Germany:
Advantages
• The short-term gap in labour is filled. Many migrants are highly skilled, eg engineers and academics. (1 mark)
• Migrants will take jobs that locals did not want and will work for lower, more competitive wages, thus reducing labour costs. (1 mark)
• Migrants will enrich the culture of the area that they move to with language, food and music. (1 mark)
• The increased population will result in an increase in the tax paid to the government, which can be invested in improving local services. (1 mark)

Disadvantages
• Migrant workers may feel discriminated against. Unemployment rises for local people. (1 mark)
• Ghettos may develop in parts of cities and there may be a shortage of affordable housing. (1 mark)
• Cost of providing services for migrant population and their families will increase, eg for schooling, healthcare, etc. (1 mark) |
| 3. | | Answers will depend on the case study referenced by the candidate.

Marks may be awarded as follows:

• For 1 mark, candidates may give one detailed explanation, or a limited description/explanation of two factors. Detail may include relevant exemplification of a problem.

• Candidate responses may be a mixture of the two styles of writing, but a maximum of 2 marks should be awarded for answers entirely consisting of limited descriptive/explanatory points.

A maximum of 2 marks should be awarded for answers which are vague or over-generalised. | 4 | *Possible answers might include:*

Eden Foundation in Nigeria:
• Educated farmers to grow perennial plants to protect the soil against heavy rain. (1 mark)
• They prevent rainsplash from dislodging fine particles and bind the loose soil. (1 mark)
• Farmers produced twice as much millet (drought-tolerant crop) as those who did not use this technique. (1 mark)
• Undisturbed by ploughing, the soil structure will remain intact. (1 mark) Organic matter holds the soil particles together.
• Stone lines are commonly used in Burkina Faso and Niger, to trap soil and water, and slow run-off. (1 mark)
• Instead, water will sink into the soil through the cracks and pours, preventing erosion. (1 mark)
• Strips of vegetation can also be used in a similar way, and can provide fodder for animals (eg Makarikari grass) or a cash crop of pumpkins could be grown. (1 mark)
• Build wells to allow effective irrigation. (1 mark)
• Contour ridges slow run-off and catch sediment before it is washed away. (1 mark) |

| 3. | | (continued) | | **Afforestation**
• To prevent soil erosion as roots will bind the soil and hold it in place. (1 mark)
• Fanya juu terraces (popular in Makanya in north-eastern Tanzania) can be made by digging a drainage channel and throwing soil uphill to make a ridge. In drier areas, trees can be planted in the ditch, and in wetter areas on the ridge. (1 mark)
• In Makanya, maize is grown between the trenches. Maize crops have increased from 1.5 tonnes per hectare to 2.4 tonnes per hectare. (1 mark) |

Section 3: Global Issues

River Basin Management

Question	General Marking Instructions for this type of question	Max mark	Specific Marking Instructions for this question
1. (a)	Candidates must explain the need for water management in Ghana using the relevant prompts from the resources. No marks should be awarded for purely descriptive points, eg 'very low rainfall in Ghana during the winter months'. Marks may be awarded as follows: • For 1 mark, candidates may give one detailed explanation, or a limited description/explanation of two factors. • Candidate responses may be a mixture of the two styles of writing, but a maximum of 2 marks should be awarded for answers which are purely descriptive.	5	*Acceptable reasons might include:* • Very low rainfall in the north of Ghana from November to March means that water management is needed to ensure that people have water for washing/drinking, etc, all year round. (1 mark) • High rainfall in the summer months throughout Ghana shows that flooding is a threat. Building dams to allow flood control is required. (1 mark) • The rapid population growth predicted in the future for Ghana suggests that demand for water from the growing population will be high and increasing (for all the usual uses of water), and this means water management is required. (1 mark) • Most people in Ghana (56%) are employed in agriculture and this shows that water management is important for irrigation purposes. (1 mark) • Only 45% of people in Ghana have access to electricity, and as their respective population grows the importance of hydroelectric power is evident. (1 mark) • Ghana's capital city of Accra is not situated on the main river (the Volta) and so it needs to be managed to allow it to benefit from domestic and industrial water use. (1 mark)
(b)	Answers must discuss the possible negative consequences. No marks should be awarded for positive consequences. Although no mark is to be awarded for the named water management project, vague/generic answers that do not relate to a specific named water management project should get a maximum of 4 marks. Candidates who only deal with socio-economic or environmental impacts should be awarded a maximum of 4 marks. For 1 mark, candidates may give one detailed explanation, or a limited description/explanation of two factors.	5	*Answers will depend on the water management project chosen, but for Ghana, possible answers might include:* **Environmental consequences could include:** • Flooding of animal habitat, eg 21% of the Bui National Park has been flooded, with fears that the rare black hippopotamus may have been threatened due to insufficient suitable habitats near the inundated area. (1 mark) • The moderation of river flow downstream of the dams could adversely affect fish habitats — 46 species of fish could be adversely affected by the Bui Dam changing the water flow, temperature and turbidity and blocking migration routes. (1 mark) • Rotting vegetation in the new lakes may release greenhouse gases, which could increase climate change. (1 mark)

1.	(b)	*(continued)*		**Socio-economic consequences could include:**
		Candidate responses may be a mixture of the two styles of writing, but a maximum of 2 marks should be awarded for answers which are purely descriptive. 1 mark can be awarded where candidates refer to **two** specific named examples *within* the case study area.		• The forcible displacing of people. In building the Akosombo Dam an estimated 80,000 people were displaced and relocated into resettlement villages (1 mark) but many of these villages were not capable of providing to the same level of income as villages previously had (with poorer soils). (1 mark) • The increased spread of diseases (eg Schistosomiasis and malaria), which have been linked to the creation of the stagnant lake. (1 mark) In addition to this, there has been an increase in the incidence of AIDS in the Volta Basin communities, linked to the increased ease of migration since the creation of Lake Volta. (1 mark) • A decline in agricultural productivity in the area surrounding Lake Volta as the soils here are less fertile than the soils now submerged under the lake. (1 mark) In addition, without the natural river floods to replace nutrients, there has been increased chemical use and the lake is now suffering from eutrophication. (1 mark) This invasion of river weeds is making fishing and navigation by motor boat across Lake Volta more difficult. (1 mark)

Development and Health

Question	General Marking Instructions for this type of question	Max mark	Specific Marking Instructions for this question
2.	Candidates may choose to answer parts (a) and (b) separately or together. Award a maximum of 6 marks for either section. Candidates must explain how each method actually helps to control the disease and not just describe or list different methods, eg releasing natural predators into the environment, such as Nile tilapia which eat the larvae. Evaluation points should also be developed points for a mark to be awarded, eg is relatively cheap, therefore affordable for developing world countries. Each evaluation should only be credited once – ie candidates should be credited for, eg, cost only once. Care should be taken not to credit reversals. For 1 mark, candidates may give one detailed explanation, or a limited description/explanation of two factors. Candidate responses may be a mixture of the two styles of writing, but a maximum of 2 marks should be awarded for answers which are purely descriptive. 1 mark can be awarded where candidates refer to **two** specific named examples *within* the case study area, up to a maximum of 2 marks.	10	*Possible answers for methods might include:* • The female anopheles mosquito acts as a vector for the transmission of malaria, so one method used was to spray pesticides/insecticides such as DDT in an attempt to kill the mosquitoes by destroying their nervous systems. (1 mark) • Breeding genetically modified sterile mosquitoes and mercenary male mosquitoes were also attempts to kill off the mosquito for good, and so stop the spreading of the disease. (1 mark) • Another method was to use specially designed mosquito traps, which mimic animals and humans by emitting a small amount of carbon dioxide in order to lure the mosquitoes into the trap where they are killed. (1 mark) • BTI bacteria can be artificially grown in coconuts and then, when the coconuts are split open and placed in a stagnant pond, the larvae eat the bacteria which destroy the larvae stomach lining, killing them. (1 mark) • Putting larvae-eating fish such as the muddy loach into stagnant ponds or paddi fields can also help to reduce the larvae as the fish eat the larvae. (1 mark) • Other methods were aimed at getting rid of the stagnant water required for mosquitoes to lay their eggs, eg draining stagnant ponds or swamps every seven days as it takes longer than this period of time for the larvae to develop into adult mosquitoes. (1 mark) • Planting eucalyptus trees, which soak up excess moisture in marshy areas, was also an attempt to prevent the formation of stagnant pools. (1 mark) • Covering water storage cans/small ponds was also used as an attempt to stop mosquitoes from reproducing successfully. (1 mark) • The increased use of insecticide-coated mosquito nets at night was an attempt to stop the mosquitoes from biting people and passing on the disease as they slept. (1 mark)

| 2. | | *(continued)* | | • Attempts were also made to cure people once they had contracted the disease by killing the plasmodium parasite once people had been contaminated with it. Drugs like Quinine, Chloroquine, Larium and Malarone were all developed in an attempt to kill the parasite. (1 mark)
• A drug developed from the Chinese herb Artemisia, and an artificial version of this called 'Oz', appears to work in some parts of the world at least, by reacting violently with the iron in the parasite and killing it before the parasite can adapt. (1 mark)

Possible comments on the effectiveness might include:

• Insecticides to kill the mosquito were effective at first and helped to eradicate the disease in Southern Europe and Florida, however the mosquito became resistant to DDT and alternative insecticides are often too expensive for developing countries. (1 mark)
• Mosquito traps have been effective at a small scale, but mosquitoes breed so quickly that it is impossible to trap them all. (1 mark)
• The approaches aimed at killing the mosquito larvae have had only limited success (and only at a local scale) and have been criticised for causing pollution/changing the ecosystem of water courses. (1 mark)
• The BTI bacteria in coconuts is a cheap and environmentally friendly solution, with 2/3 coconuts clearing a typical pond of mosquito larvae for 45 days. (1 mark)
• Draining stagnant ponds is impossible to be effective on a large-scale, especially in tropical climates where it can rain heavily most days. (1 mark)
• Using mosquito nets at night/covering up exposed skin is effective as mosquitoes are often most active during dusk and dawn. (1 mark)
• Drugs to kill the parasite once inside humans have been effective for a spell, but the parasite often adapts and becomes resistant — this is true even of the Artemisia-based drugs in SE Asia. (1 mark)
• Anti-malarial drugs often have unpleasant side-effects such as nausea, headaches and in some cases hallucinations. (1 mark)
• They are also expensive to research, develop and produce, making them often too expensive for people living in developing countries. (1 mark)
• Attempts are ongoing to develop a vaccine that could eradicate malaria for good, but so far this has not been successful. (1 mark) |

Global Climate Change

Question		General Marking Instructions for this type of question	Max mark	Specific Marking Instructions for this question
3.	(a)	Award a maximum of 1 mark for an explanation of the greenhouse effect. 1 mark should be awarded for a source along with an explanation of why it is increasing, eg methane is released from cattle's digestive system and beef is increasingly in demand across the world. A maximum of 2 marks should be awarded for a list of sources of individual gases.	5	*Possible answers might include:* • Carbon dioxide from burning fossil fuels — road transport, power stations, heating systems, cement production — and from deforestation, particularly in the rainforests where more carbon dioxide is present in the atmosphere and less being recycled in photosynthesis (1 mark) and peat bog reclamation/development (particularly in Ireland and Scotland for wind farms). (1 mark) • CFCs: disused refrigerators release CFCs when the foam insulation inside them is shredded. (1 mark) The coolants used in fridges and air conditioning systems create CFCs which are safe in a closed system, but can be released if appliances are not disposed of correctly. (1 mark)

Question	General Marking Instructions for this type of question	Max mark	Specific Marking Instructions for this question

3. (a) (continued)

- Methane: from rice paddies to feed rapidly increasing populations in Asian countries such as India and China (1 mark), belching cows to meet increasing global demand for beef. (1 mark) Methane released from permafrost melting in Arctic areas due to global warming. (1 mark)
- Nitrous oxides: from vehicle exhausts and power stations. (1 mark)
- Sulphate aerosol particles and aircraft contrails: global 'dimming' — increase in cloud formation increases reflection/absorption in the atmosphere and therefore cooling. (1 mark)

3. (b) — Max mark: 5

For 1 mark, candidates may give one detailed explanation, or a limited description/explanation of two factors.

Candidate responses may be a mixture of the two styles of writing, but a maximum of 2 marks should be awarded for answers which are purely descriptive.

1 mark can be awarded where candidates refer to **two** specific named examples (species, ocean currents, or use of numeracy) *within* the case study area.

Possible answers might include:

- Rise in sea levels caused by an expansion of the sea as it becomes warmer and also by the melting of glaciers and ice caps in Greenland, Antarctica, etc. (1 mark)
- Low-lying coastal areas, eg Bangladesh affected with large-scale displacement of people and loss of land for farming and destruction of property. (1 mark)
- More extreme and more variable weather, including floods, droughts, hurricanes, tornadoes becoming more frequent and intense. (1 mark)
- Globally, an increase in precipitation, particularly in the winter in northern countries such as Scotland, but some areas like the USA Great Plains may experience drier conditions. (1 mark)
- Increase in extent of tropical diseases, eg yellow fever as warmer areas expand, possibly up to 40 million more in Africa being exposed to risk of contracting malaria. (1 mark)
- Longer growing seasons in many areas in northern Europe for example, increasing food production and range of crops being grown. (1 mark)
- Impact on wildlife, eg extinction of at least 10% of land species and coral reefs suffer 80% bleaching. (1 mark)
- Changes to ocean current circulation, eg in the Atlantic the thermohaline circulation starts to lose impact on north-western Europe, resulting in considerably colder winters. (1 mark)
- Changes in atmospheric patterns linking to monsoon, El Nino, La Nina, etc. (1 mark)

Trade, Aid and Geopolitics

Question	General Marking Instructions for this type of question	Max mark	Specific Marking Instructions for this question
4. (a)	Candidates should explain the inequalities in world trade patterns. For 1 mark, candidates may give one detailed explanation, or a limited description/explanation of two factors. Candidate responses may be a mixture of the two styles of writing, but a maximum of 3 marks should be awarded for answers which are purely descriptive.	5	*Possible answers might include:* - Developing countries often sell primary products at low value, therefore profits are limited. (1 mark) - Often, many countries are producing the same raw material, which keeps prices low. (1 mark) - However, developed countries manufacture products, which adds value and provides increased profits. (1 mark) - Developing countries are prevented from setting up coffee processing plants as high import taxes would be placed on processed coffee, whilst developed countries import coffee beans. (1 mark) - Patterns established during colonial times have been difficult to break. (1 mark) - Limits and quotas are also enforced, eg Kenya's export of coffee to the European Union is subject to a tariff of 9%, whilst other countries are subject to a 3.1% tariff. (1 mark)

4.	(a)	*(continued)*		• Developing countries are often very dependent on one or two products, eg bananas, sugar or copper. (1 mark) • Developed countries set the prices for raw materials through trading on commodity exchanges around the world, eg the New York Mercantile Exchange. (1 mark)
	(b)	Candidates should explain the effectiveness of strategies to reduce inequalities in world trade. For 1 mark, candidates may give one detailed explanation, or a limited description/explanation of two factors. Candidate responses may be a mixture of the two styles of writing, but a maximum of 2 marks should be awarded for answers which are purely descriptive. 1 mark can be awarded where candidates refer to **two** specific named examples *within* the case study area.		*Possible answers might include:* • The World Trade Organisation, established in 1996, settles trade disputes and continues to promote free trade and the removal of tariffs and quotas. (1 mark) • The removal of trade barriers means that developing countries will have access to lucrative markets in the developed world. (1 mark) • Some countries, eg in the Caribbean, are attempting to diversify their trade by developing non-traditional exports such as new crops or manufactured goods. Others are pursuing new markets. (1 mark) • The creation of trade alliances: the Caribbean Community and Common Market (CARICOM) was established to promote trade between Caribbean countries. (1 mark) • Customs duties between member states were removed, thus even the smallest Caribbean countries have access to a regional market. (1 mark) • Member countries are encouraged to purchase raw materials from other CARICOM countries. This has spread the benefits of industrialisation and has encouraged industries to locate in the smaller countries. (1 mark) • Within CARICOM, the Organisation of Eastern Caribbean States (OECS) has been established, which groups together the seven smallest countries in terms of their population. (1 mark) • OECS has created a single currency which makes trade between OECS countries much easier as money is not lost in transactions. (1 mark) • The OECS developed a common agricultural policy to subsidise farmers and removed controls on the movement of workers, allowing skilled workers to migrate. (1 mark) • Fairtrade guarantees a fair price for produce which always covers the cost of production regardless of the market price. (1 mark). • Five-year rolling contracts can be given which allows long-term planning to take place for investment in farm machinery, education, etc. (1 mark)

Energy

Question		General Marking Instructions for this type of question	Max mark	Specific Marking Instructions for this question
5.	(a)	No marks for describing the differences — candidates must account for the differences.	4	Although at present the amount of energy used by 'developed countries' is slightly higher than the amount used by 'developing countries', this is forecast to change. In the next 20 years the energy use in 'developing countries' is expected to increase at a much faster rate, whereas the 'developed countries' rate is expected to be stable, only rising slowly. Why is this? *Possible answers might include:* • Most of the global economic and population growth is happening in the 'developing' countries. (1 mark) This is causing an increase in demand from: o **Residential use** — with increased prosperity comes an increase in the standards of living for billions of people. Electricity for lighting and appliances such as televisions, washing machines, air conditioning, etc, will all cause energy use to increase. (1 mark)

5.	(a)	*(continued)*		<ul><li>○ **Industrial use** — unlike in 'developed' countries, much of the economic growth in 'developing' countries is based on energy-hungry manufacturing industries. This accounts for some of the increased energy use here. (1 mark)</li><li>○ **Transport** — in a global economy many of the manufactured products are sold to 'developed' countries, and therefore need to be transported around the world — using energy. (1 mark) As people in 'developing' countries become more prosperous, car ownership rates will also increase, causing more energy use. (1 mark)</li><li>In 'developed' countries the population growth rates are more stable (or even declining), and so there is not any great increase in demand for energy. (1 mark) New products and technologies are also increasingly more energy-efficient which keeps energy consumption here more steady. (1 mark)</li></ul>
	(b)	Candidates must comment on the suitability of each renewable approach discussed. Care should be taken not to credit reverse statements twice (eg solar energy is effective in Spain because it has long hours of sunshine, whereas it is less effective in Scotland where there are fewer hours of sunshine). For 1 mark, candidates may give one detailed explanation, or a limited description/explanation of two factors. Candidate responses may be a mixture of the two styles of writing, but a maximum of 3 marks should be awarded for answers which are purely descriptive. 1 mark can be awarded where candidates refer to **two** specific named relevant examples.	6	*Possible answers might include:* <ul><li>**Hydroelectric power** is more effective where there is high rainfall to ensure that reservoirs are always at capacity (1 mark), and suitable underlying geology (impermeable rock) to ensure water is not lost from the reservoirs through seepage. (1 mark) Hanging valleys often make ideal sites for effective hydroelectric power because they allow the vertical drop of water needed to power turbines. (1 mark) Hydroelectric pump storage schemes allow electricity production to be instantly produced as required, and are currently being used to meet periods of peak demand in Scotland. (1 mark)</li><li>**Wind power** is most effective where there are no barriers to the prevailing wind to allow regular and reliable movement of air to turn the turbines. (1 mark) Concerns have been raised about how to bridge the energy gap when the wind turbines are not generating electricity on calm days, because it is difficult to store the electricity produced from wind turbines. (1 mark) Other drawbacks to this approach, for developing countries, focus on the initial high costs of construction, but after this the energy produced is cheap. (1 mark)</li><li>**Wave power** approaches are currently being developed, and are most effective in areas such as the Pentland Firth where the fetch is large giving powerful waves, and in areas such as Cornwall.</li><li>**Tidal power** is used in areas where there is a large tidal range to create tidal currents which can be used to turn turbines, eg in the Pentland Firth or Bay of Fundi.</li><li>**Solar energy** is most effective where there are long hours of intense sunshine (eg in Spain) to power the solar panels. (1 mark)</li><li>**Geothermal energy** is most effective in tectonically active zones (such as Iceland), where there is a heat source (from magma) closer to the surface of the Earth, which can be used to generate steam. (1 mark) It is a reliable source of energy and can be used as needed. (1 mark)</li><li>**Biomass energy and biofuels** can provide continuous energy as required by the burning of plant matter. (1 mark) Drawbacks include concerns about air pollution to the local area increasing, and using land that is needed for crops production in developing countries. (1 mark) As carbon dioxide released equals what the plants recently took in, biomass energy does not add new greenhouse gasses, and so is more environmentally friendly than burning fossil fuels. (1 mark)</li></ul>

Section 4: Application of Geographical Skills

Question	General Marking Instructions for this type of question	Max mark	Specific Marking Instructions for this question
1.	Candidates should make reference to all sources, including the OS map to evaluate the suitability of the route in relation to the brief. For 1 mark, candidates should refer to the resource and offer an explanation with reference to the brief, or a limited description/explanation of two factors. A maximum of 5 marks should be awarded for answers consisting solely of limited descriptive points. A maximum of 4 marks should be awarded for candidates who give vague over-generalised answers which make no reference to the map. There are a variety of ways for candidates to give map evidence including descriptions, grid references and place names. A maximum of 4 marks should be awarded for answers which make no evaluation of the plan.	10	*Possible answers might include:* **Suitable for all levels of runners:** • The roads and streets are narrower here, causing a 'bunch' start for the runners which may cause problems. (1 mark) • While the steep uphill section on the A15 and even steeper downhill section on the B1188 may cause some difficultly for runners (1 mark), the generally flat nature of the route is likely to allow for a quick time, encouraging runners. (1 mark) • The exposed area between the 4km and 7km marks may cause some difficulties with the weather conditions/wind. (1 mark) **Cause minimum disruption to people and business in the local area:** • In grid square 9769 and 9770 the route goes through a residential area — local people may be unhappy with the lack of access by road. (1 mark) • The litter/noise levels (bands on the run) from runners/spectators may also cause disruption in residential areas. (1 mark) • The route uses the A57 (dual carriageway), a major road — closure will cause increased congestion and disruption. (1 mark) • Having the run on a Sunday will minimise this disruption as the area/businesses are likely to be quieter. (1 mark) **Promote business in local area:** • The route starts and finishes in the CBD/historic centre of Lincoln, promoting the local area to runners/supporters. (1 mark) • The estimated number of runners with extra supporters could be beneficial for some shops near to the start/finish line. (1 mark) • Hotels and camp/caravan sites will have extra business due to the increase in visitors. (1 mark) • Holding the run in February would bring in visitors at a time when trade is traditionally low. (1 mark) **Provide access for runners to start line:** • Car parking and transport to the start line may be problematic due to the lack of space in the CBD. (1 mark) • The start/finish of the route is near to the train and bus station allowing for some of the runners to get there by public transport and avoid parking. (1 mark) **Scenic/interesting for runners:** • The start/finish is in the historic centre of Lincoln (Castle/Cathedral), which may be of interest to the runners. (1 mark) • There is a varied landscape going through rural and urban environments, which may be more scenic for the runners. (1 mark) **Improvements may include:** • Starting the race in Hartsholme country park/West Common race course is likely to allow for easier access by car/parking. (1 mark) Hartsholme country park would be more scenic, as there is more green space and lakes. Having the run in the summer months may attract a bigger number of runners/spectators and boost trade. (1 mark)

General Marking Information for students

When answering a question, you should ensure that you read the question carefully before you start. You should look at the command word to ensure you answer the question correctly, eg if the question asks you to explain, then you must give reasons to support your response. Remember that there are no describe questions so, if your answers do not contain more than a descriptive point, it will be difficult to achieve any marks. If a question asks for two points of view to be covered, for example, advantages and disadvantages, then both must be covered for full marks. If the question is out of six and only the advantages are covered, then a maximum mark of four or five might be awarded. If a named area or example is asked for, then you may lose a mark(s) if you give a general response to the question. There are often more points covered (worth a mark each) in the following answers than will be required. For example, there could be six or seven points listed for a five-mark answer. Since the question is only worth five marks then any five of the six points could gain the five marks.

HIGHER GEOGRAPHY 2015

Section 1: Physical Environments

Question	General marking principle for this type of question	Max mark	Specific Marking Instruction for this question
1.	Award 1 mark for a developed explanation, two limited explanations, or for a description of the discharge with limited explanation. Award a maximum of 2 marks for description of the discharge with 1 mark being awarded for every two descriptive points being made. Markers should accept all relevant and appropriate explanations for the source provided.	4	Slight increase in discharge until 09.00 hours in response to rain which started to fall at 07.00 hours (**1 mark**). At first, this rain would have been intercepted by vegetation and have infiltrated the soil (**1 mark**). There is a steep rising limb up to a peak discharge of 100 cumecs at 18.00 hours (**1 mark**). This water would have filled up storages in the soil due to throughflow and groundwater (**1 mark**) — as the soil became saturated, surface run-off increased causing a peak (**1 mark**). The rising limb becomes less steep briefly between 13.00 and 15.00 hours, caused by a marked reduction in rain to 4mm around 10.00 hours (**1 mark**). There is a short lag time of 5 hours which could be due to deforestation/steep slopes/impermeable rock (**1 mark**). A high number of tributaries may lead to the short lag time as water is transported more rapidly by surface run-off (**1 mark**). The river discharge quickly decreases, shown by a steep falling/recession limb down to because of no more rainfall after 15.00 hours (**1 mark**).
2.	Award a maximum of 4 marks for either feature. Award a maximum of 6 marks if no diagrams are used. Check any diagram(s) for relevant points not present in the text and award accordingly. Well-annotated diagrams that explain conditions and processes can gain full marks. Award a maximum of 1 mark for three or more correctly named, but undeveloped, processes. Award a maximum of 2 marks for fully developed processes for any one feature. Answers which are purely descriptive, or have no mention of any processes or conditions, should achieve no more than 2 marks in total, with 1 mark being awarded for every two descriptive points being made. **1 mark** Limited explanation — the use of the names of at least two processes in context with no development of these. **2 marks** The use of the names of at least two processes with development of these, but no other reference to conditions. **OR** Limited use of the names of at least two processes, with at least two descriptive points about the landscape formation.	7	Headland with weaknesses such as joints, faults or cracks is eroded by the sea to form firstly caves (**1 mark**). Erosion takes place due to hydraulic action — pounding waves compress trapped air in the rocks, creating an explosive blast which weakens and loosens rock fragments (**1 mark**), abrasion/corrasion — rock fragments thrown against the headland create a sandblasting (abrasive) action, wearing away the rock (**1 mark**), solution/corrosion — carbonic acid in sea water weathering limestone and chalk (**1 mark**), attrition — rock fragments slowly being ground down by friction from wave action into smaller and rounder pieces (**1 mark**). In some cases, a blowhole can form in the roof of the cave as compressed air is pushed upwards by the power of the waves, causing vertical erosion (**1 mark**). Over time, erosion on both sides of the headland cuts through the backwall and enlarges the cave to create an arch (**1 mark**). Continued erosion at the foot of the headland and the effects of vibrations on the roof of the arch weakens it, eventually resulting in the collapse of the arch roof, leaving a stack isolated from the headland (**1 mark**). These low ridges of sand or shingle slowly extend from the shore across a bay or a river estuary and are caused by longshore drift (**1 mark**). This lateral movement occurs when waves, driven by the prevailing wind, pushes material up the beach; known as the swash (**1 mark**). The returning backwash is dragged back by gravity down the beach at right angles (**1 mark**). Material slowly builds up to appear above the water and begins to grow longer and wider. The spit develops as long as the supply of deposits is greater than the amount of erosion (**1 mark**). Spits form when there is a change in direction on a coastline, which allows a sheltered area for deposition (**1 mark**). They can also develop at a bay or a river estuary where the river current prevents the spit from extending right across the bay or estuary (**1 mark**). The shape can change through time to become curved or hooked at the end in response to changes in wind direction and currents (**1 mark**).

Question		General marking principle for this type of question	Max mark	Specific Marking Instruction for this question
2.		*(continued)*		
		3 marks Two developed processes with limited explanation of how the feature forms over time.		
		4 marks Two named processes with development of these, with two further statements explaining the formation of the feature.		
3.		Award 1 mark for each developed explanation or for two less developed points. Candidates should be awarded 1 mark if they only name and locate at least two cells/winds, not credited elsewhere. Credit any valid responses.	4	Warm air rises at the Equator and travels in the upper atmosphere to around 30^0 N and S, cools and sinks **(1 mark)**. Air moves from the tropical high to the low pressure area at the equator creating the Hadley cell/ Trade Winds **(1 mark)**. Cold air sinking at the poles moves to 60°N/S to form the Polar Cell/Polar Easterlies **(1 mark)**. The cold air from the poles meets warmer air from the tropics, causing air to rise creating the Ferrel Cell low pressure **(1 mark)**. Air is moved from the tropical HP, towards LP at the polar front, forming the westerlies **(1 mark)**. This convergence causes the air to rise, with some of this flowing in the upper atmosphere to the Poles where it sinks, forming the Polar cell. Easterly winds blow away from the high pressure at the Pole. **(1 mark)**. Warm air from the Equator is distributed to higher and cooler latitudes and cold air from the Poles distributed to lower and warmer latitudes **(1 mark)**. Due to the Coriolis effect winds are deflected to the right in the northern hemisphere **(1 mark)**. Credit should be awarded for answers which refer to the Rossby Waves and/or the Jet Stream.

Section 2: Human Environments

Question		General marking principle for this type of question	Max mark	Specific Marking Instruction for this question
4.		Candidates should discuss the positive or negative consequences of the predicted population structure in 2050. For 1 mark, candidates may give one detailed consequence, or a limited description/explanation of two factors. Detail may include relevant exemplification of a problem. A maximum of 2 marks should be awarded for answers consisting entirely of limited descriptive/explanatory points, with two such points required for 1 mark. Care should be taken not to ensure consequences are relevant to developing countries. Credit any other valid responses.	5	*Possible answers may include:* • The total population will increase significantly putting additional pressure on services and resources like education **(1 mark)**. Housing in Ghana, like many developing countries, is already overcrowded. This problem is likely to continue, with many people being forced to live in shanty town housing **(1 mark)**. • There will be a much larger potential workforce which may attract multinational companies to the country **(1 mark)**. An increase in the active age group, however, could also result in higher levels of unemployment or underemployment **(1 mark)**. • In total, there will be twice as many children, so significant investment in maternity hospitals, immunisation programmes and education will be needed **(1 mark)**. It will be necessary to build more schools and train more teachers to support the growing number of young people **(1 mark)**. • Government policies may promote smaller families or encourage emigration to reduce the problems of over-population **(1 mark)**. • With life expectancy increasing, it will also be necessary to invest in health-care to meet the needs of an ageing population in the future **(1 mark)**.

Question	General marking principle for this type of question	Max mark	Specific Marking Instruction for this question
5.	Answers will depend on the case study referenced by the candidate. Marks may be awarded as follows: For 1 mark, candidates may give one detailed explanation, or a limited description/explanation of two factors. A maximum of 2 marks should be awarded for answers consisting entirely of limited descriptive points, with two such points required for 1 mark. A maximum of 4 marks should be awarded if the answer does not clearly relate to a specific case study. Credit any other valid responses.	5	*For candidates who write about Glasgow, possible answers might include:* • Bypass/ring road: In Glasgow, the M74, M77 or Glasgow Southern Orbital Route mean that through traffic does not need to travel into the city centre **(1 mark)**. • Pedestrianised area in centre: Sections of Argyle Street and Buchanan Street were pedestrianised, the main shopping streets are no longer congested to make it safer and more pleasant for shoppers **(1 mark)**. • Park and ride and improvements to public transport: Commuters have been encouraged to travel into the city centre by train, underground or bus to reduce the number of cars on the road **(1 mark)**. Additional parking has been provided at suburban train stations, eg Hamilton West and Carluke, and new railway stations have opened, eg Larkhall **(1 mark)**. The introduction of bus lanes has reduced journey times along main commuter routes, therefore encouraging people to travel by bus **(1 mark)**. • One way streets: To improve the flow of traffic, the streets around George Square and Hope Street are now one way **(1 mark)**. • Parking restrictions and fines: Parking charges have been increased (60p for 12 minutes in the city centre, with a maximum stay of 2 hours) and traffic wardens patrol the streets to discourage people from bringing their cars into the centre (£60 fines are issued as necessary) **(1 mark)**. • Multi-storey car parks: Multi-storey car parks have been built, eg Buchanan Galleries, to reduce the number of cars parking on the surrounding streets **(1 mark)**. • Bridges and tunnels: Bridges like the Clyde Arc ('Squinty Bridge') and the Clyde Tunnel have been built to improve the flow of traffic from the north to the south of the city, diverting from the CBD thus reducing congestion **(1 mark)**.
6.	Answers will depend on the case study referenced by the candidate. Marks may be awarded as follows: For 1 mark, candidates should briefly describe a strategy and offer one evaluative point. Further developed/detailed evaluative comments should be awarded 1 mark each. At least two strategies are required for full credit. A maximum of 3 marks should be awarded for answers consisting entirely of limited evaluative points. Up to 2 marks can be awarded for description or explanation of strategies (i.e. no evaluation), with two such points required for 1 mark. Credit any other valid responses.	5	*Possible answers might include:* • **Large scale redevelopment:** The Dharavi Redevelopment Project where local people will be moved to high rise apartment blocks, however, only those who have been resident in Dharavi since 2000 will be eligible to move into these apartments **(1 mark)**. Other residents will be moved to other parts of the city, which will break up communities and may result in people being too far from their work **(1 mark)**. The new flats will also be too small for those who currently have workshops above their homes **(1 mark)**. • **Slum Rehabilitation:** has planned and managed improvements such as upgrading mains sewerage to help reduce diseases such as cholera however, within twelve years, only 15% of Dharavi was redeveloped **(1 mark)**. • **Local projects:** Self-help schemes support the efforts of local people to improve their housing for example by adding an additional floor to buildings thus reducing overcrowding **(1 mark)**. Toilets have been added and are shared by two or three families who help to keep them clean, which has reduced the incidence of water related diseases **(1 mark)**.

Section 3: Global Issues

Question		General marking principle for this type of question	Max mark	Specific Marking Instruction for this question
7.	(a)	1 mark should be awarded for each detailed explanation. Award a maximum of 4 marks if only human or physical factors are explained (although a physical factor could be linked to a human issue, for example narrow cross section which would reduce the cost). A maximum of 2 marks should be awarded for answers consisting entirely of limited descriptive points, with two such points required for 1 mark. Credit any other valid responses.	5	Narrow cross section of the valley is required in order to reduce construction costs of the dam (1 mark). A deep valley is required behind the dam as this will result in a smaller surface area for the reservoir, thereby reduce loss from evaporation (1 mark). A site which has impermeable rock would be advantageous as this would reduce loss from the reservoir by percolation (1 mark). A site which is close to construction materials would help to reduce the cost of transporting these materials to the construction site (1 mark). An area free from earthquakes or subsidence is needed as the area needs to be able to support the weight of a large dam (1 mark). A site close to areas of farmland or urban areas would help to reduce water/electrical loss during transportation (1 mark). The costs involved in moving people who live in the area to be flooded, to reduce costs for compensation and re-housing (1 mark). The land which is to be flooded should not be valuable, for example high quality farmland or of historic/environmental importance (1 mark). An area with plentiful supply of snowmelt/rainfall/river water is required to ensure a consistently high volume of water in the reservoir (1 mark).
	(b)	Award 1 mark for each detailed explanation. Candidate answers must include both social and economic consequences. No marks should be awarded for negative impacts. Award a maximum of 4 marks if the answer does not clearly relate to a specific named water management project. A maximum of 2 marks should be awarded for answers consisting entirely of limited descriptive points, with two such points required for 1 mark. Credit any other valid responses.	5	**Answers will depend on the water management project chosen but for the Aswan High Dam, possible answers might include:** • Increased access to clean drinking water reduces water borne diseases such as typhoid (1 mark). • Increased irrigation, (with 33,600km^2 of irrigated land) which allows for two crops a year to be grown, reducing malnutrition (1 mark) production of wheat and sugar cane tripled allowing more export crops to be produced (1 mark). • Increase in hydroelectric power (from the 12 generating units in the Dam, these generate approx. 2.1 gigawatts) attracting industries such as smelting industries (1 mark). • The introduction of the Nile perch and tiger fish into Lake Nasser has increased the commercial fishing industry and fishing tourism industry (1 mark). • Industries which require large amounts of water have grown up near to Aswan, for example the Egyptian chemical industry KIMA which makes fertilisers (1 mark).
8.	(a)	1 mark should be awarded for each detailed comparison, or a comparison with a short explanation. A detailed comparison will contain a qualitative statement such as dramatically higher and be supported by the statistics. A maximum of 2 marks should be awarded for answers which are purely descriptive and do not go beyond making comparisons directly from the table, with two such comparisons required for 1 mark. Where candidates refer only to the headings (i.e. not to the data), a maximum of 1 mark should be awarded. Markers should take care to look for comparisons wherever they occur in a candidate's answer. Credit any other valid responses.	4	GDP per person shows that Brazil, Mexico and Cuba are emergent developing countries with an intermediate level of income, while Kenya and Malawi are still classed as low income developing countries (1 mark). Employment in agriculture is far higher in Malawi (90%) than Brazil at 16% so it is less industrialised (1 mark). Adult literacy is far higher in Brazil, Mexico and Cuba than in Malawi suggesting they have more schools (1 mark). Birth rate shows a vast difference in development with both African countries having high birth rates (Kenya — 30%); compared with Mexico at 18%, suggesting lack of contraception/education (1 mark). There is also a large difference between the country with the poorest life — Malawi has a life expectancy of 53 whereas Cuba is 78, suggesting better health care (1 mark).

Question		General marking principle for this type of question	Max mark	Specific Marking Instruction for this question
	(b)	1 mark should be awarded for each detailed explanation, or for two more straightforward explanations A maximum of 2 marks should be awarded for four straightforward descriptive lists exemplified by country name, eg Nigeria can sell oil to make money. There is no need to mention the countries in the question and other countries can be used to exemplify points made. Credit any other valid responses.	6	Some countries have natural resources such as oil, which can be sold to generate foreign currency (1 mark). Some countries eg Chad are landlocked, and find it more expensive to export and import goods (1 mark). Countries with a poor education system have many low skilled workers and are unable to attract foreign investment (1 mark). Countries with fertile soils and a suitable climate can grow cash crops which can be sold for income (1 mark). Corruption in government such as in Nigeria can lead to money being used inappropriately (1 mark). Where countries suffer from conflict or civil war they are unable to keep the economy working and spend extra finance on weapons (1 mark). Countries which have accumulated large debts have to repay loans and interest causing less money for services (1 mark). Famine can lead to malnutrition, and a reduced capacity to work and create income (1 mark).
9.	(a)	No marks should be awarded for naming the greenhouse gases as these are included in the diagram. No marks should be awarded for explanations of the enhanced greenhouse effect as this is included in the diagram. 1 mark should be awarded for each detailed explanation, or for two undeveloped points. Undeveloped points may include comparative potency of each gas, or limited reasons for increase. Markers should take care not to credit the same reason twice, eg car ownership. Credit any other valid responses.	4	*Possible answers might include:* The Enhanced Greenhouse effect has been caused by an increase in greenhouse gases within the atmosphere. **Carbon Dioxide (remains in the atmosphere for 100 years):** • Burning fossil fuels, for example coal, oil and natural gas release carbon dioxide into the atmosphere, which will trap heat. Coal has been used increasingly to power factories, generate electricity in power stations and to heat homes (1 mark). • Increased car ownership has resulted in more petrol and diesel being used to fuel cars (1 mark). • Deforestation, especially in the Amazon Rainforest, has resulted in less carbon dioxide being absorbed, and the burning releases more CO_2 (1 mark). • Peat bog reclamation (particularly in Scotland and Ireland) during, for example, the construction of wind farms, has also resulted in additional carbon dioxide being released into the atmosphere (1 mark). **Methane:** (More than 20 times as effective in trapping heat than CO_2; accounts for 20% of the enhanced greenhouse effect; remains in the atmosphere for 11–12 years) (1 mark). • Methane has been released from landfill sites as waste decomposes and when drilling for natural gas (1 mark). • The increase in padi fields to feed rapidly growing populations in Asian countries has increased the amount of methane in the atmosphere (1 mark). • The increasing demand for beef has resulted in more methane being created by belching cattle and from animal dung (1 mark). **Nitrous oxide:** • Nitrous oxide is 200–300 times more effective in trapping heat than carbon dioxide. Increased car exhaust emissions have resulted in more nitrous oxide (1 mark). • Due to rising food demand the increased production of fertilisers also adds to the amount of nitrous oxide in the atmosphere (1 mark).

Question		General marking principle for this type of question	Max mark	Specific Marking Instruction for this question
9.	(a)	*(continued)*		**Chlorofluorocarbons (CFCs):** • Refrigerators which are not disposed of correctly release CFCs when the foam insulation inside them is shredded **(1 mark)**. • Air-conditioning — the coolants used in air conditioning systems create CFCs, which must be disposed of correctly **(1 mark)**. Credit any relevant developed points.
	(b)	For 1 mark, candidates should give one detailed explanation, or a limited description/explanation of two factors. A maximum of 3 marks should be awarded for answers consisting entirely of limited descriptive points, with two such points required for 1 mark. A maximum of 2 marks should be awarded for answers consist of limited description with a specific example. 1 mark can be awarded where candidates refer to **two** specific named examples accurately linked with an effect (species, ocean currents, or use of figures). Candidates should be credited for both positive and negative effects. Credit any other valid responses.	6	*Possible answers might include:* **Global Effects:** • Sea level rises caused by an expansion of the sea as it becomes warmer and also by the melting of glaciers and ice caps in Greenland, Antarctica, etc **(1 mark)**. • Low-lying coastal areas, eg Bangladesh affected with large-scale displacement of people and loss of land for farming and destruction of property **(1 mark)**. • More extreme and more variable weather, including floods, droughts, hurricanes, tornadoes becoming more frequent and intense **(1 mark)**. • Globally, an increase in precipitation, particularly in the winter in northern countries such as Scotland, but some areas like the USA Great Plains may experience drier conditions **(1 mark)**. • Increase in extent of tropical/vector borne diseases, eg yellow fever as warmer areas expand, possibly up to 40 million more in Africa being exposed to risk of contracting malaria **(1 mark)**. • Longer growing seasons in many areas in northern Europe for example, increasing food production and range of crops being grown **(1 mark)**. • Impact on wildlife and natural habitats, eg extinction of at least 10% of land species and coral reefs suffer 80% bleaching **(1 mark)**. • Changes to ocean current circulation, eg in the Atlantic the thermohaline circulation starts to lose impact on north—western Europe, resulting in considerably colder winters **(1 mark)**. • Changes in atmospheric patterns linking to changes in the monsoon caused by El Nino, La Nina **(1 mark)**. • Increased risk of forest fires, for example in Australia and California due to change in surface temperatures and changes in rainfall patterns **(1 mark)**.
10.	(a)	Candidates should refer to both social and economic impacts for full credit, although markers should be aware that many impacts could be considered both social and economic. 1 mark should be awarded for each detailed explanation. A maximum of 3 marks should be awarded for answers consisting entirely of limited points, with two such points required for one mark. A developed point may be a detailed explanation of an impact, or may be two less detailed impacts. Candidates may choose to answer a mixture of two styles. Credit any other valid responses.	4	*Possible answers might include:* • Economic impacts include many people being trapped in poverty and trying to survive on very little money. This might be because companies in the developed world want to manufacture their product for the cheapest price possible **(1 mark)**. • Many countries are unable to make a decent profit on the goods they sell because they are forced to pay tariffs to "developed countries" in trading blocs **(1 mark)**. • Government subsidies and grants in "developed countries" allow companies to sell products (eg rice and grain) at a cheaper price than is possible in many "developing countries" — undercutting local farmers and causing them to lose money **(1 mark)**. • Social impacts of this poverty include many children working instead of going to school, and this causes an illiterate population, with the consequent lack of opportunities and poorer quality of living for people that this applies to **(1 mark)**.

Question		General marking principle for this type of question	Max mark	Specific Marking Instruction for this question
10.	(a)	*(continued)*		• Creates a self-perpetuating cycle of poverty where the next generation are unable to access well paid employment due to being illiterate **(1 mark)**. • The lack of well-paid jobs means many people live in shanty town type accommodation, with little access to clean water, safe electricity, sanitation, etc **(1 mark)**.
	(b)	Using the information in Diagram Q10b, and also your knowledge of fair trade, explain how fair trade: (i) helps to reduce inequalities in world trade, and; (ii) impacts on farming communities. 1 mark should be awarded for each developed point. A maximum of 4 marks should be awarded for answers consisting entirely of limited points, with two such points required for 1 mark. A developed point may be a detailed explanation or a description of a trend with a less detailed explanation or may be two less detailed explanations. Credit any other valid responses.	6	*Possible answers might include:* • The graph shows that the price of normal coffee is very variable — for example when there was a drought in Brazil, the global production decreased and so the price of coffee increased **(1 mark)**. • When China and India increased demand this increased competition for coffee and caused the price to increase **(1 mark)**. • When the production of coffee increased this eased competition for coffee and caused the price to drop **(1 mark)**. • In all these cases, fair trade matched any price increases, but also never dropped below 125 cents per kilo. This allows farmers benefiting from fair trade to have security for the future and to make plans to improve their quality of life **(1 mark)**. • Fair Trade Coffee also does not lose money to the extra "links" in the chain that normal coffee has. Each link in the chain (for example the "middle men" companies), demand a fee and take money away from the farmer. This leaves the farmer with almost nothing to show for growing the coffee in the first place **(1 mark)**. • Fair Trade also ensures that the environment of the coffee plantations is not misused and damaged for future generations by farming in a sustainable way **(1 mark)**. • Fair Trade also ensures that there are no child workers, so that children get the chance to attend school and improve their opportunities in the future **(1 mark)**. • Fair Trade not only benefits the individual farmers, as there are community improvement payments to villages supplying fair trade coffee beans. This allows for example, new schools and health centres to be built and staffed that would otherwise not have been built **(1 mark)**.
11.	(a)	1 mark should be awarded for each developed point for energy production or for two undeveloped points. A developed point may include a descriptive statistic/comparative statement with explanation or a developed explanation. Where candidates have only described the data, award a maximum of 1 mark. 1 mark should be awarded for two descriptive points. Credit any other valid responses.	4	*Possible answers might include:* • Saudi Arabia gets 100% of its energy from fossil fuels because it has vast natural resources of oil and gas in its territory **(1 mark)**. • Portugal has the largest amount of other renewables because its climate is sunny, with many cloud-free days for solar power **(1 mark)** and windy with mid-latitude westerly winds blowing over Atlantic Ocean to Portugal for wind power **(1 mark)**. • France has the largest amount of nuclear energy because it has no oil or gas reserves and has used up most of its coal reserves **(1 mark)**. France used to import oil and gas, but started to develop its own nuclear power in a big way after the "oil shock" — when OPEC countries suddenly decided to quadruple the price for oil in 1973 **(1 mark)**. • Paraguay has the largest amount of hydroelectric power because it has large amounts of tropical precipitation and major rivers that can be harnessed **(1 mark)**. • Iceland has the largest amount of geothermal energy because it is located at a plate boundary where the heat of magma is closer to the surface and so can be harnessed efficiently **(1 mark)**.

Question	General marking principle for this type of question	Max mark	Specific Marking Instruction for this question
(b)	1 mark should be awarded for each developed explanatory point or for two undeveloped points. Take care not to credit reverse statements (for example renewable sources do not pollute and non-renewable sources pollute). Maximum of 4 marks for one energy source. Credit any other valid responses.	6	*Possible answers for all renewable energy sources might include:* • It harnesses a free resource, which is sustainable for the future **(1 mark)**. It does not contribute greenhouse gases such as CO_2 or other pollution in to the atmosphere **(1 mark)**. **Also, for:** • Wind power can be very effective if sited properly in an exposed area where there is regular wind flow to turn the blades **(1 mark)**. A large are of open land is required, close to large cities, for both wind and solar to avoid loss when transporting **(1 mark)**. However, when the wind is not blowing the turbines cannot generate any energy, and at present it is very difficult to store wind power to use when demand needs it **(1 mark)**. Wind turbines built on peat bogs require the bogs to be drained — which releases carbon dioxide that would otherwise have been stored — causing pollution **(1 mark)**. • Solar power is effective on clear, sunny days but is far less effective when clouds block direct sunlight **(1 mark)**. Solar power can be stored in batteries and used at times of high demand **(1 mark)**. Photostatic panels and solar panels have a very high initial cost. **(1 mark)**. • Hydroelectric power can be generated as required and so is able to cope with peaks and troughs in demand **(1 mark)**. • Geothermal power can be generated as required and so is able to cope with peaks and troughs in demand **(1 mark)**. It is most effective in areas close to plate boundaries where the heat source underground (magma) is closer to the surface **(1 mark)**. • Tidal power has the potential to generate vast quantities of energy. Eg experimental tidal generators in the Pentland Firth **(1 mark)**. Tidal power, unlike wind/solar is predictable, therefore reliable. **(1 mark)**. • Technology for wave power, is not yet sufficiently advanced to allow for large scale economic production. **(1 mark)**. *Possible answers for all non-renewable energy sources might include:* It can generate power as required and so is able to cope with peaks and troughs in demand **(1 mark)**. However, it harnesses a resource, which will eventually run out and then will effectively be gone for good, so is not a viable long-term energy solution **(1 mark)**. Coal, gas and oil contribute towards global warming and cause air pollution **(1 mark)**. **Also for:** • Coal/gas are a relatively cheap way for a country to generate energy. They can also be moved about relatively easily to where power is being generated **(1 mark)**. • Gas causes less pollution than burning coal (up to 45% less) or oil (up to 30% less) **(1 mark)**.

Question		General marking principle for this type of question	Max mark	Specific Marking Instruction for this question
11.	(b)	*(continued)*		• Nuclear — due to small amounts of radioactive material needed to produce huge amounts of energy resources will last longer (**1 mark**). Nuclear power does not produce greenhouse gases such as CO_2 or other forms of air pollution (**1 mark**). However, it does produce nuclear waste, which is toxic and needs to be carefully stored for hundreds of thousands of years (**1 mark**). Also, nuclear reactors are expensive to build and run so are not suitable for all countries (**1 mark**). A radioactive leak could be devastating for the local people and environment and for future generations (**1 mark**). • Technology in many developed countries is geared towards non-renewable sources as power stations/infrastructure are already in place (**1 mark**). • Supply of oil may be disrupted and process rise due to sabotage during wars (**1 mark**).
12.		Candidates should make reference to all sources, including the OS map to discuss the suitability and impact of the centre. 1 mark should be awarded for each developed explanatory point. 1 mark should be awarded where candidates refer to the resource and offer a brief explanation of its significance (beyond the wording of the resource), or give a limited description/explanation of two factors. A maximum of 5 marks should be awarded for answers consisting entirely of limited descriptive points, with two such points required for 1 mark. A maximum of 4 marks should be awarded for candidates who give vague over-generalised answers, which make no reference to the map. There are a variety of ways for candidates to give map evidence including descriptions, grid references and place names. It is possible that some points referred to as a disadvantage may be interpreted by other candidates as a negative impact. Markers should take care to credit each point only once, where it is best explained.	10	*For suitability, possible answers may include the following positive observations:* **Possible advantages of this location** The surrounding area offers potential for a range of activities such as hillwalking, with steep slopes posing a challenge for mountain biking eg Auchenroy Hill (**1 mark**), there are numerous trails in nearby woodland which could be used for orienteering eg Galloway Forest Park 488038 (**1 mark**). Steep, sheer cliffs in the disused quarry at 457054 could be used for abseiling/climbing (**1 mark**). There are a number of opportunities for watersports such as the River Doon could be used for fishing, or the River Doon and Bogton Loch/Loch Doon could be used for kayaking (**1 mark**). Local schools at 477060 may be interested in using the facilities for outdoor learning and enriching the curriculum. The SSSI at Bogton Loch and the Nature Reserve at Dalmellington Moss would offer opportunities for exploring the wildlife present (**1 mark**). Variety of attractive scenery including Dalcairnie Linn waterfall offering potential for geography studies (as per spec) at 467043 (**1 mark**). There are a number of tourist facilities already in the area (Industrial Railway Centre, Horse Riding, Fishing), and these could work together with the new facility to join up and benefit from each other's custom (**1 mark**). The centre would be only 0.5 km from the A713 main road and there is access to the site at present, with the B741 to the west and a minor road lying to the east (**1 mark**). **Possible disadvantages of this location** Large area of marsh/poorly drained land close to the site would present major construction difficulties and could result in expensive remediation work (**1 mark**). There could be a serious threat of flooding from the loch overflowing after heavy precipitation (**1 mark**). **Positive impacts** Creation of jobs during construction and in running the centre would be welcome in East Ayrshire where unemployment is about 1–2% above the national average, as shown in Table Q12 (**1 mark**).

Question	General marking principle for this type of question	Max mark	Specific Marking Instruction for this question
12.	*(continued)*		This could boost the local economy by attracting more visitors because East Ayrshire at 18% has the lowest income from tourism in Ayrshire and Arran, as shown in Diagram Q12 (b) **(1 mark)**. This could help to reverse rural depopulation in the area, as Diagram Q12 (a) shows that there was a decrease of about 1,000 people in Dalmellington and Burnton between 2001 and 2010 **(1 mark)**.
			Negative impacts Construction work and the various activities could be very damaging to the fragile environment and could face strong opposition from conservationists **(1 mark)**. The centre and associated infrastructure could spoil the views as it is on flat land, very visible from the road **(1 mark)** and there may be increased air and noise pollution traffic to the area harming the fragile ecosystems in the SSSI **(1 mark)**. Increased litter could lead to farm animals being harmed which could mean a financial loss for nearby farmers — Craigengillan Home Farm 475038 **(1 mark)**.

HIGHER GEOGRAPHY
2016

Section 1: Physical Environments

Question			General marking principle for this type of question	Max mark	Specific Marking Instructions for this question
1.	(a) (b)		1 mark should be awarded for each developed comparison, or two limited comparisons up to a maximum of 3 marks. For full marks, candidates must answer both parts of the question. Award **one mark for each detailed explanation or for two limited explanations.** Candidates may choose to answer this question holistically and should be credited accordingly. Award 1 mark for a limited comparison with a limited explanation. Credit any other valid responses	5	**Comparison** should highlight the marked contrast in precipitation totals, seasonal distribution and number of rain days. Figures may provide detail for a comparison, however they are not required. **Answers may include:** Agadez has 200mm rainfall compared with Lagos 1600mm. Agadez has a peak in August, whereas at Lagos there is a higher peak June. **(1 mark)** Agadez has a distinct dry season from October to May with one peak, whereas Lagos has year round rainfall with 2 peaks. **(1 mark)** **Explanation** should focus on explaining the migration of the ITCZ and the movement of the Maritime Tropical and Continental Tropical air masses over the course of the year. **Answers may include:** Lagos sits south of the ITCZ and is influenced by hot, humid maritime tropical air from the Gulf of Guinea for most of the year. **(1 mark)** The twin precipitation peaks can be attributed to the ITCZ moving northwards in the early part of the year and then southwards later in the year in line with thermal equator/overhead sun. **(1 mark)** Agadez, on the other hand, is under the influence of hot, dry continental tropical air blowing from the Sahara and lies well to the north of the ITCZ for most of the year. **(1 mark)**
2.	(a)		1 mark should be awarded for each developed explanation, or for two limited descriptions/explanations. If candidates discuss more than one glaciated area, mark all and award marks to the highest scoring section. If candidates discuss a coastal area, bracket the name off, and award marks to any conflict/strategy which could apply to a glaciated area. Bracket off responses referring to social/economic conflicts, although be aware that candidates may develop this by referring to economic consequences. Credit any other valid responses.	5	Precise points will depend on the conflict and area chosen (although the extracts refers to tourism, note that the question is open to all environmental conflicts). Traffic congestion on narrow rural roads leads to high levels of air and noise pollution. **(1 mark)** Tourists parking on grass verges in honeypot locations such as Bowness can lead to erosion of fragile grass verges. **(1 mark)** Tourists wander off footpaths widening them and stone wall can be damaged by people climbing over them. **(1 mark)** Litter causes visual pollution and can harm wildlife (or livestock) if it is eaten. **(1 mark)** Speedboats on lakes produce oil pollution and can erode beaches. **(1 mark)** Quarrying such as Honister Quarry in the Lake District can produce large quantities of dust which can settle on plants stunting their growth. **(1 mark)** Large lorries travelling to and from the quarries can cause structural damage due to vibrations from the heavily loaded vehicles. **(1 mark)**

Question			General marking principle for this type of question	Max mark	Specific Marking Instructions for this question
	(b)	(i) (ii)	1 mark should be awarded for each developed explanation and each developed evaluation, or for two limited descriptions/explanations/evaluations. Award 1 mark for a limited explanation with a limited evaluation. Candidates should be awarded a maximum of 4 marks if there is no evaluation. Candidates may choose to answer this question holistically and should be credited accordingly.	5	Removing litter bins in remote areas where it is difficult to empty them (leading to overflowing bins), encourages people to take their litter home. **(1 mark)** Traffic restrictions such as one way streets and limited waiting times have had limited success as people prefer the convenience of their own vehicles. **(1 mark)** Using farmers' fields as temporary car parks reduces on-street parking and can bring in another form of income for the farmer. **(1 mark)** Hosing lorries or covering with tarpaulin has reduced the amount of dust, and transporting by train takes lorries off the road. **(1 mark)** Planting trees around unsightly developments can shield them, but this is a long term solution. **(1 mark)** New developments are controlled by NPA by-laws ensuring they use local materials which blend in with the landscape. **(1 mark)** Speed limits to reduce beach erosion have been implemented, however, this has resulted in speedboat users moving to other lakes. **(1 mark)**

Section 2: Human Environments

Question		General marking principle for this type of question	Max mark	Specific Marking Instructions for this question
3.	(a) (b)	Answers will depend on the case study referenced by the candidate. For full marks candidates must answer both parts of the question. Marks may be awarded as follows: For 1 mark, candidates may give one detailed explanation, or a limited description/explanation of two factors. For 1 mark, candidates may give a developed comment on effectiveness or two limited comments on effectiveness. For 1 mark, candidates may give one limited explanation with one limited comment on effectiveness. Candidate responses are likely to address the question holistically and marks should be awarded accordingly. Credit any other valid responses.	6	*Suitable methods for Africa include:* • Afforestation projects reduce wind erosion and prevent soil erosion as the tree roots bind the soil and hold it in place **(1 mark)**. • However, the current and anticipated cost of the Great Green Wall project has been heavily criticised; it is out of the reach of most developing nations **(1 mark)**. • Fanya juu terraces (popular in Makanya in north-eastern Tanzania) have been made by digging a drainage channel and throwing soil uphill to make a ridge to increase infiltration **(1 mark)**. • This low technology approach has been particularly successful. However, maintaining the terraces is very labour intensive **(1 mark)**. • In Makanya, maize is grown between the trenches increasing crop yield (from 1.5 tonnes per hectare to 2.4 tonnes per hectare.) which reduces the need to cultivate more marginal farmland **(1 mark)**. • Diguettes or "Magic Stones" are lines of stones placed along the contours of gently sloping land to trap rain water as well as soil **(1 mark)**. • This is particularly useful following the seasonal rainfall in the Sahel caused by the ITCZ which causes surface run-off **(1 mark)**. • By preserving the most fertile top-soil, stone lines have increased yields by 40% in some areas **(1 mark)**.

Question		General marking principle for this type of question	Max mark	Specific Marking Instructions for this question
		(continued)		• Zai (microbasins, or planting pits) are hollows dug to retain moisture and nutrients (**1 mark**). • This ensures year round plant coverage which increases infiltration and reduced run-off (**1 mark**). *Other methods may include:* • Microdams • Education programmes • Managed grazing/moveable fencing *NB Suitable methods for a rainforest area may include:* • Re afforestation • Increased global co-operation • Crop rotation • Agroforestry • National Parks
4.	(a)	For 1 mark, candidates may give one detailed explanation, or a limited description/explanation of two factors. Credit any other valid responses.	3	Population data can be gathered by: • Census is a survey carried out every ten years to gather population data (**1 mark**). • Census: Each householder is asked to complete a detailed questionnaire about the number of people living in their home, their age, gender, employment, home and languages spoken (**1 mark**). • Civil registrations of births, marriages and deaths keep an up to date count of the population (**1 mark**). • Sampling – Population surveys are conducted to gather social and economic data, and can be conducted at regional, national or international levels (**1 mark**). • In China National Population Sample Surveys have been conducted annually, with 1% of the population being asked to complete the form (**1 mark**). • Government records: information on migration may be gathered from visa applications or Borders Agency (**1 mark**). • Data from electoral roll and NHS records allows population data to be updated in between census collection (**1 mark**).
	(b)	Candidates must explain the problems of collecting accurate population data in developing countries. For 1 mark, candidates may give one detailed explanation, or a limited explanation of two factors. Named examples will enhance a candidate's answer, however, a name alone will not gain any credit. Credit any other valid responses.	6	*Problems of gathering population data:* • Language barriers – countries with many official languages have to translate their census forms and employ enumerators who can speak multiple languages (**1 mark**). • Literacy levels – many people can't read and write, and therefore are unable to complete the forms, or might make mistakes unintentionally (**1 mark**). • Size of the population: the sheer size of some populations make it very difficult to conduct a census, eg in China and India (**1 mark**). • Inaccessibility: The poor infrastructure and difficult terrain, for example in the Amazon Rainforest, may make it difficult for enumerators to distribute census forms (**1 mark**). • Wars/civil wars: Conflict can make it too dangerous for enumerators to enter, or for data to quickly become dated (**1 mark**). • Cost: Undertaking the census is a very expensive process, even for developed world countries. In developing countries, there may be higher priorities for spending, including housing, education and health care (**1 mark**).

Question		General marking principle for this type of question	Max mark	Specific Marking Instructions for this question
		(continued)		• Migration: Rapid rural to urban migration can make it difficult to gather accurate population data as data will become outdated very quickly (**1 mark**). • Many people in developing countries may be living in shanty towns, eg Dhararvi, or are homeless, so have no official address making it difficult to count them (**1 mark**). • People who are illegal immigrants are unlikely to complete a census for fear of deportation, leading to inaccurate data (**1 mark**). • Nomadic people: Large numbers of migrants, eg the Tuareg, Fulani or Bedouin, or shifting cultivators in the Amazon can easily be missed or counted twice (**1 mark**). • Ethnic tensions and internal political rivalries may lead to inaccuracies, eg northern Nigeria was reported to have inflated its population figures to secure increased political representation (**1 mark**).

Section 3: Global Issues

Question		General marking principle for this type of question	Max mark	Specific Marking Instructions for this question
5.	(a)	1 mark should be awarded for a detailed explanation or a limited description/explanation of two factors; this may include the use of facts from the resources. Credit any other valid responses.	6	• Population increase of almost 50 million for country in last 20 years would require additional water for domestic use (**1 mark**). • Irrigation is required for rice production to feed the population or for export (**1 mark**). • Textile industries use large volumes of water therefore a reliable year round supply is required (**1 mark**). • Due to the monsoon climate a lack of rainfall in Nov – March increases the need for water to be stored to allow use during the dry period (**1 mark**). • The heavy monsoon rainfall of up to 800mm of rain in the month of July means there is a requirement to prevent flooding (**1 mark**). • The city of Chittagong has a population of 7.5 million and is situated on the banks of the river; this increases the need for flood prevention (**1 mark**). • Only 62% of the country has access to electricity, HEP from the dam could be used to improve this (**1 mark**). • Excess energy produced could be exported to neighbouring countries such as India (**1 mark**). • Improved sanitation means that far less of the population will be at risk from diseases such as cholera (**1 mark**).
	(b)	Answers must discuss the possible positive **and** negative **environmental** impacts. Award a maximum of 3 marks for either. 1 mark should be awarded for a developed explanation, or a more straightforward explanation linked to the case study. Care should be taken not to credit purely social or economic benefits but markers should be aware that some candidates will be able to link these to the environment. Award a maximum of 3 marks if the answer is vague/does not relate to a specific named water management project.	4	*Answers will depend on the water management project chosen but for the Aswan High Dam, possible answers might include:* • Lake Nasser provides a sanctuary for waterfowl and wading birds and has more than 32 species of fish (**1 mark**). • River and irrigation water becomes saline with high evaporation rates resulting in farmers downstream having to switch to more salt-tolerant crops (**1 mark**). • The change in river regime has caused the loss of many animal habitats eg the drying up of the Nile delta area may lead to inundation of sea water (**1 mark**).

Question			General marking principle for this type of question	Max mark	Specific Marking Instructions for this question
	(b)		*(continued)*		• The water table is rising in the Nile valley, which is resulting in major erosion of foundations of ancient temples and monuments such as Abu Simbel **(1 mark)**. • Increase in 'clean' hydro-electric power from the 12 generating units in the Dam, instead of using polluting fossil fuels **(1 mark)**. • The lack of flooding and subsequent lack of silt deposition has led to a need for chemical fertilisers which has resulted in high levels of Nitrogen and Phosphorous being washed into rivers **(1 mark)**. • The sediments which were transported to the river mouth forming a delta are now trapped behind the dam, a situation which has led to severe erosion along the Egyptian coast **(1 mark)**.
6.	(a)		1 mark should be awarded for each detailed evaluation of the indicator, or for two more straightforward reasons. Candidates may respond by explaining the benefits of composite or other indicators. This should be credited accordingly.	3	• Single indicators such as GNI are averages which can hide extremes within a country such as a rich minority and a poor majority **(1 mark)**. • Development is not only about money, other aspects of development like literacy and healthcare are also important **(1 mark)**. • Composite indicators such as HDI give a more rounded picture; by combining a number of indicators **(1 mark)**. • It is not possible to tell where a rise in GNI is being spent within a country—not always spent on improving standard of living **(1 mark)**. • GNI is always expressed as US$ to allow comparison, however exchange rates continually fluctuate **(1 mark)**. • GNI does not take into account the informal economy however this accounts for a large proportion of wealth generated in some countries **(1 mark)**.
	(b)	(i) (ii)	1 mark should be awarded for each developed explanation or for two less developed explanations. A developed point may be: • a detailed explanation • a description of a strategy with a less detailed explanation • two less detailed explanations Evaluations may refer to why a strategy is suited to a developing country or by giving data to show how successful a programme has been. Answers which provide no evaluation should be awarded a maximum of 6 marks. Candidates may choose to answer holistically and should be credited accordingly. 1 mark should be awarded for a limited explanation with a limited evaluation. Candidate may discuss malaria in this question; only strategies which could be considered PHC should be credited.	7	• Oral Rehydration Therapy is the mixture of salt and sugar with clean water to help people suffering from diarrhoea **(1 mark)**; it is very effective as it is cheap and simple and it can be administered by untrained staff **(1 mark)**. • Vaccination programmes such as the UNICEF run polio immunisation campaign were delivered to rural areas as people here find it more difficult to access healthcare **(1 mark)**. • By 2014 polio was endemic in only 3 countries (Afghanistan, Pakistan and Nigeria) **(1 mark)**. • Charities such as Water Aid work with countries and other aid agencies to improve water and sanitation by installing e.g. pit latrines **(1 mark)**. • By 2010 the number of people without access to improved drinking water had decreased to (11%) and the ash compost from latrines can improve crop yield **(1 mark)**. • Barefoot Doctors provide health education through play and songs as many people are illiterate in developing countries **(1 mark)**. • Insecticide treated bed nets provide a physical barrier against the mosquito and kills the mosquito preventing further spread **(1 mark)**. • However, they need to be treated regularly to be effective and in some cases are damaged by being used as fishing nets **(1 mark)**. • Play Pumps International provide roundabouts which extract ground water which can be used for drinking **(1 mark)** these provide local people with transferable skills and use appropriate level of technology **(1 mark)**.

	Question		General marking principle for this type of question	Max mark	Specific Marking Instructions for this question
7.	(a)		1 mark should be awarded for each detailed explanation or for a limited description/ explanation of two factors. Markers should take care not to credit human causes of climate change. Credit any other valid responses.	4	*Possible answers might include:* • Milankovitch's theory: changes in the earth's orbit and tilt alter the amount of energy reaching the Earth **(1 mark)**. • Every 41,000 years, there is a change in the tilt of the Earth's axis. A greater tilt means more sunlight in polar regions **(1 mark)**. • Over a 97,000 year cycle, the Earth's orbit stretches, affecting the amount of energy received **(1 mark)**. • Sunspot activity: global mean temperatures can be raised by peaks of sunspot activity, which follow an 11 year pattern **(1 mark)**. • Volcanic eruptions: After violent eruptions, large amounts of dust and droplets of sulphur may reflect the sun's rays lowering temperature **(1 mark)**. • Retreating ice caps release additional fresh water leading to changes in oceanic circulation **(1 mark)**. • This also reduces the albedo effect as reflection has decreased as more land is exposed **(1 mark)**. • Melting Permafrost: Methane being released from melting permafrost from decomposing organic matter **(1 mark)**.
	(b)		For 1 mark, candidates should give one detailed explanation of the strategies used to manage climate change, or a limited description/explanation of two factors. Candidates do not need to refer to local, national and international strategies to gain full credit. Named examples will enhance a candidate's answer; however an example alone will not gain any credit. Credit any other valid responses.		**Local—** • Individuals can reduce, reuse and recycle products so that less refuse is sent to landfill sites. This will reduce the amount of methane entering the atmosphere **(1 mark)**. • To reduce the amount of carbon dioxide generated by the burning of fossil fuels, households could reduce energy consumption by insulating their homes or switching lights off, etc. **(1 mark)**. • People could also be encouraged to use public transport, walk or cycle, or use hybrid or electric cars to cut down on fossil fuel consumption **(1 mark)**. • Fridge disposal should be managed carefully to ensure CFC gases don't escape. New cooling units no longer emit CFC's **(1 mark)**. **National—** • Government Policies such as 'Helping Households to cut their Energy Bills' encourage the use of 'Smart Meters' improving energy efficiency **(1 mark)**. • Increasing the use of low carbon technologies such as windfarms – the UK Government is committed to creating 15% of energy by renewable sources **(1 mark)**. **International—** • The Paris agreement outlined agreements between leaders of developed and developing countries to limit climate change below a 2° rise **(1 mark)**. • The European Union has committed to reducing carbon emissions by 20% by 2020. The EU will reward developing countries financially **(1 mark)**. • The impact of climate change could also be managed by preparing for extreme weather events, for example, flood defences could be built to hold back flood water, or flood plains and natural wetlands could be used to store flood water **(1 mark)**.

	Question		General marking principle for this type of question	Max mark	Specific Marking Instructions for this question
8.	(a)		1 mark should be awarded for each detailed comparison, or a comparison with a short explanation. A detailed comparison will contain a qualitative statement such as dramatically higher and be supported by the statistics. A maximum of 2 marks should be awarded for answers which are purely descriptive and do not go beyond making comparisons directly from the table, with two such comparisons required for one mark. Where candidates refer only to the headings (ie not to the data), a maximum of one mark should be awarded. Markers should take care to look for comparisons wherever they occur in a candidate's answer. Credit any other valid responses.	4	*Possible answers might include:* • Some countries can afford to buy in lots of imports (like USA who spend $2,273 billion) whereas other countries cannot afford to (like Zimbabwe who spend $4 billion) **(1 mark)**. • Some countries can sell manufactured goods to make money (for example China who make $2,210 billion from exports), whereas other countries rely on raw materials (for example Botswana who make $3 billion from exports) **(1 mark)**. • China has the biggest trade surplus of $438 billion compared to the USA, which has the biggest trade deficit of $698 billion **(1 mark)**. • Two countries with similar population sizes are Australia and Ghana, yet Australia has a trade surplus of $7 billion and a GDP per capita of $67,304, whereas Ghana has a trade deficit of $5 billion and a GDP per capita of $3,500 **(1 mark)**. • Two countries with similar population sizes are China and India, yet China has a trade surplus of $438 billion and a GDP per capita of $6,071, whereas India has a trade deficit of $198 billion and a GDP per capita of $1,499 **(1 mark)**.
	(b)		1 mark should be awarded for each detailed explanation, or for two more straightforward explanations. A maximum of 2 marks should be awarded for 4 straightforward descriptive lists, exemplified by country names. Take care not to credit the reverse points of answers with regards to primary/ manufactured products. Credit any other valid responses.	6	*Possible answers might include:* • Some countries have the knowledge and technology to make and sell manufactured products, which sell at high prices and so larger profits are made **(1 mark)**. • These manufactured products sell for a more stable price so countries can plan for the future with confidence in future income **(1 mark)**. • These primary products sell for prices that fluctuate, and so countries cannot invest in development for the future **(1 mark)**. • Often it is the 'developed countries', which set the price for primary products and keep them as low as possible by playing 'developing countries' against each other **(1 mark)**. • Some countries are too reliant on one or two low value exports, and so if anything happens to the price/production of these export products, the country's economy is hit badly **(1 mark)**. • For example, Saudi Arabia exports oil which is in high demand, whereas Burkina Faso exports shea nuts **(1 mark)**. • Trading Blocs (like the EU) can control the trade terms for the benefit of its members and make it difficult for non-members to do as well **(1 mark)**. For example, they can set up tariffs and import duties that they charge non-member countries, which makes their goods appear less competitive **(1 mark)**. • They can also set quotas, which put a limit on the amount of product that a non-member country can sell to the member country **(1 mark)**.

Question			General marking principle for this type of question	Max mark	Specific Marking Instructions for this question
9.	(a)		1 mark should be awarded for each developed point or for two less developed points. A developed point may be a detailed explanation or a description of a trend with a less detailed explanation or may be two less detailed explanations. Credit any other valid responses.	5	• Increased vehicle ownership due to 2 car households therefore increased demand for petrol (1 mark). • Increased ownership of electronic devices such as tablets, due to changing technology and affordability therefore increased demand for electricity (1 mark). • Increased standard of living and more single occupancy households leading to more houses with central heating systems (1 mark). • Improved energy efficiency in residential sector – for example energy-saving fridges and LED lighting (1 mark). • Improved insulation of housing such as cavity wall insulation cuts down on heat loss causing less heating to be required (1 mark). • Improved efficiency in cars and the growth of more affordable fuel efficient, 'greener' hybrid cars (1 mark). • Government initiatives such as the cycle to work scheme encourages people to leave their cars at home by subsiding the cost of cycle purchase (1 mark). • The Government signed up to the Kyoto Protocol to reduce greenhouse gas emissions. This has led to targets for industry to meet in terms of energy savings (1 mark). • Significant dip around 2008 due to declining industrial output caused by the recession and subsequent lowering of manufacturing industry in the UK (1 mark).
	(b)		Award 1 mark for each developed advantage or disadvantage or for every two undeveloped points. Candidates must discuss advantages and disadvantages to gain full credit. Candidates must discuss a non-renewable source of energy. No marks for discussing renewable sources of energy. Candidates are expected to consider different aspects of their chosen source of energy. Credit any other valid responses.	5	*Possible answers for all non-renewable energy sources might include:* • They provide instant power as required meeting demand at peak times such as early evening (1 mark). • They cause air pollution and release greenhouse gases so contribute towards global warming (1 mark). *For 'fracking' other possible answers could include:* • The shale gas provides an alternative energy source reducing reliance on traditional fossil fuels such as oil which are finite. (1 mark). • Noise and light pollution is increased due to 24hr production on shale gas sites (1 mark). • In USA shale gas production has allowed it to become self-sufficient in gas and means it does not have to rely on imports from other countries (1 mark). • However, the fracking fluid used in the process could pollute ground water and enter the domestic water system (1 mark). • The fracking process could be linked to causing minor earthquakes and tremors in the local area leading to structural damage to buildings and infrastructure (1 mark).

Section 4: Application of Geographical Skills

Question		General marking principle for this type of question	Max mark	Specific Marking Instructions for this question
10.	(a)	Candidates should make reference to all sources, including the OS map to discuss the suitability and impact of the by-pass route. It is possible that some points referred to as a disadvantage may be interpreted by other candidates as a negative impact. Markers should take care to credit each point only once, where it is best explained. 1 mark should be awarded where candidates refer to the resource and offer a brief explanation of its significance, or give a limited description/explanation of two factors. A maximum of 5 marks should be awarded for answers consisting solely of limited descriptive points with 2 such points required for 1 mark. A maximum of 4 marks should be awarded for candidates who give vague over-generalised answers, which make no reference to the map. There are a variety of ways for candidates to give map evidence including descriptions, grid references and place names. Credit any other valid responses	10	*Possible advantages of this route:* • For the first 700m of the route at 745094, the road will be following a dismantled railway line. This will make the road easier to build here, and so reduce costs (**1 mark**). *Possible disadvantages of this route:* • The new road will require several bridges to be built over rivers and the railway line (774113), which will increase the costs (**1 mark**). • A road cutting required on approach to Upper Wilting farm as land rises steeply from Combe Haven may require more complex engineering works (**1 mark**). • The new road crosses an area of marshland in square 7510, which will require costly drainage (**1 mark**). • The new road is in the flood plain of Watermill Stream/ Combe Haven, and so is at risk of flooding (**1 mark**). *Possible impacts on the surrounding area might include:* **Negative Environmental Impacts** Diagram Q10C makes it clear that Sussex Wildlife Trust think the road will cause "unacceptable environmental damage". They might be referring to the fact that: • It will require the destruction of deciduous woodland at Chapel Wood and Park Wood (7711), which will cause habitat loss (**1 mark**). • The road will spoil a natural landscape and cause visual pollution for walkers on the 1066 country walk (**1 mark**). • The route passes close to sites of conservation value (SSSI, SNCI and an AONB), which could be spoiled by the amount of noise and air pollution a new road will bring (**1 mark**). • The photograph shows street lighting following the new road which will cause light pollution to a rural area (**1 mark**). **Positive Environmental Impacts** • Graph 10C shows that Glyne Gap is currently suffering from unsafe levels of air pollution on many days (sometimes as many as 16 days in a month), which the by-pass will reduce, improving the air quality here (**1 mark**). • This will also reduce the noise pollution for local people in this area such as the suburbs of Bexhill eg Pebsham (**1 mark**). **Negative Socio-Economic Impacts** • The route passes through or very nearby several farms (Acton's Farm, Adam's Farm, Lower Wilting Farm for example), which could cause the disruption to the farmers as farm animals could be frightened by the noise of vehicles using the new road (**1 mark**). • The route passes over several footpaths and bridleways including the "1066 Country Walk – Bexhill Link", which might be closed or at least spoiled by a busy new road (**1 mark**).
				Positive Socio-Economic Impacts • Diagram Q10A shows that the A259 at Glyne Gap is currently very busy all day (with a peak flow of just under 2600 vehicles in an hour). The new road would reduce the traffic flow here dramatically, reducing traffic congestion (**1 mark**). • This would save journey/commuting time and reduce transport costs for local people/businesses (**1 mark**). • Building the new road will create many jobs and increase money for people in the local area, which will boost the economy (**1 mark**).

Acknowledgements

Permission has been sought from all relevant copyright holders and Hodder Gibson is grateful for the use of the following:

Image © Peter R Foster IDMA/Shutterstock.com (SQP page 14);
Population Pyramids for Ghana 2013 and 2050, taken from the US Census Bureau. Public domain (2015 page 5);
Congestion levels in UK Cities, as identified by TomTom © TomTom International B.V. (2015 page 6);
Image © gary yin/Shutterstock.com (2015 page 7);
The table 'Development indicators for selected developing countries' taken from the Central Intelligence Agency website (https://www.cia.gov/library/publications/the-world-factbook/index.html). Public domain (2015 page 10);
Diagrams taken from National Park Service U.S. Department of the Interior (http://www.nps.gov/goga/learn/nature/images/Greenhouse-effect.jpg). Public domain (2015 page 11);
An extract from 'Lake District Management Plan – State of the Park Report 2005: Access and Recreation' © Lake District National Park Authority (2016 page 4);
The table 'Trade Patterns of Selected Countries' taken from the CIA World Fact Book 2013 (https://www.cia.gov/library/publications/the-world-factbook/index.html). Public domain (2016 page 12);
An extract © Sussex Wildlife Trust (2016 page 14);
Two images © Department of Transport. Contains public sector information licensed under the Open Government Licence v3.0 (https://www.nationalarchives.gov.uk/doc/open-government-licence/version/3/) (2016 page 15);
Ordnance Survey maps © Crown Copyright 2016. Ordnance Survey 100047450.